INTEGRATING INFORMATION *into the* ENGINEERING DESIGN PROCESS

edited by **Michael Fosmire and David Radcliffe**

Purdue University Press, West Lafayette, Indiana

INTEGRATING INFORMATION *into the*
ENGINEERING DESIGN PROCESS

Purdue Information Literacy Handbooks

Sharon Weiner, Series Editor

CONTENTS

Clarify the Task

Synthesize Possibilities

Select Solution

Refine Solution

Communicate Effectively

Improve Processes

PART III *Ensuring That Students Develop Information Literacy Skills*

FOREWORD

There is wide recognition that information literacy is an essential element of success in academic work, employment, and everyday life. Though many variations of definitions of information literacy abound, I consider information literacy to be a way of thinking—a *habit of mind*. Its defining characteristic is the drawing upon information-related strategies and skills, almost instinctively, to address problems or questions. For students, the development of this habit occurs optimally through the integration of information literacy concepts, skills, and strategies in courses, curricula, and cocurricular activities. It becomes a habit through progressive reinforcement during the formal educational process.

There are foundational information literacy competencies that are common to most situations. There are also specialized information literacy competencies that one would apply as contexts vary. For example, information literacy in academic work differs from that in the workplace or for personal uses. Disciplines are examples of varying contexts that influence information literacy. Students and practitioners in the sciences would draw on different information skills, strategies, and resources to solve problems or answer questions than those in the humanities or social sciences. These adaptations of information literacy should be grounded within a discipline through a deep understanding of its paradigms. These include the foundational concepts, models, and pedagogies that underpin the discipline.

It is with pride that I introduce *Integrating Information into the Engineering Design Process*, the first book in the Purdue Information Literacy Handbooks series. It is an outstanding example of the application of information literacy in a discipline. No other work has so thoroughly and capably integrated information literacy with the learning of engineering design. The authors and editors have succeeded in presenting a cohesive and evidence-based approach to an engineering paradigm: the design process. Working in close collaboration, engineering faculty, staff, and information specialists have developed a groundbreaking resource.

I invite proposals for future handbooks in the Purdue Information Literacy Handbooks series, the purpose of which is to promote evidence-based practice in teaching information literacy competencies through the lens of the different academic disciplines. The handbooks will include the perspective of disciplinary experts as well as library and information science professionals. For more information, please refer to the Purdue University Press website at www.press.purdue.edu.

Sharon Weiner, EdD, MLS
Series Editor
Professor and W. Wayne Booker Chair in Information Literacy, Purdue University Libraries
Vice President, National Forum on Information Literacy

PREFACE

Our goal in creating this book was to develop something unique—to fill a gap in the resources available to engineering faculty and engineering librarians. There is a singular absence of practical advice on how to apply information literacy concepts in the domain of engineering education. For a number of years, faculty in the Libraries and in the School of Engineering Education at Purdue University have been collaborating to help first-year engineering students make more informed design decisions—decisions based on wise use of available information sources. Both engineering educators and librarians understand that novice engineering students tend to make quick decisions about what approach to take to solve a problem, then spend a lot of time developing prototypes and finishing details, when they might have saved a lot of effort and created a superior outcome had they spent more time upfront attempting to understand the problem more fully and thinking more broadly about potential solutions before actually working to implement one.

Furthermore, many engineering students seem to believe that everything needs to be done from first principles. They waste an inordinate amount of time trying to redesign a widget that is already cheaply and readily available commercially, and often spend months designing a new device, only to find out that something remarkably similar had already been patented years ago. This well-intentioned but wasted effort can be mitigated by helping engineering students adopt a more informed approach to engineering design. To date there has not been a systematic effort to develop such a model that resonates with both engineers and librarians. This book was conceived to meet that need.

Librarians and engineering educators each hold a piece of the puzzle in developing an integrated, informed learning approach, and this book is written for both audiences, as a way to bridge the gaps in conceptualization and terminology between the two important disciplines. Librarians specialize in the organization and application of information, while engineers understand not only the practice of engineering design, but also how students learn and what cognitive barriers they may have to adopting new concepts and ways of knowing. Over the past few years, the Colleges of Engineering and Technology at Purdue have, collaboratively with the engineering librarians, developed first-year courses that substantively integrate information literacy into their design activities. Our experiences in this integrated and synergistic approach are what we have endeavored to capture in this book.

We, the editors, developed and tested the central organizing principle of this book, the Information-Rich Engineering Design (I-RED) model, as the framework for integrating information literacy into a capstone design course, IDE 48500, Multidisciplinary Engineering, as part of the Multidisciplinary Engineering program at Purdue.

We approach the creation of this book as a design activity itself. A team of engineering educators, engineering librarians, and communications experts was assembled and a first prototype of the book was created at a two-day workshop held at Purdue University in September 2012. This event afforded a unique opportunity for the contributors to make suggestions about their and each other's chapters and for clarifying what content should be located in which chapter. Over the course of the writing, we also had the chance to try out each other's techniques in the classroom, providing additional feedback on the effectiveness of different activities. The result, we hope, is that even though this work was written by a collection of individual authors, both engineers and librarians, it will read as a collective, integrated whole.

Truly, it has been a pleasure to work with all the talented writers and thinkers who devoted their time to this book. We had many excellent conversations, and we, the editors, know our teaching practice has improved greatly from the exchange of ideas over the course of the writing.

INTRODUCTION

This handbook is structured in three distinct parts. Chapters 1 through 3 assemble key concepts about information literacy, engineering design and how engineers use information. These chapters draw on the relevant bodies of literature and are written in a scholarly style. Specifically, Chapter 1 views the engineering design process from several quite different perspectives. The goal is not to settle on a preferred model of design but to identify generic characteristics that are common to most normative descriptions of how design is done. Chapter 2 is an overview of concepts and definitions in information literacy, and Chapter 3 provides some evidence of what practicing engineers and engineering students actually do when carrying out design activities. Chapter 4, the final chapter in Part I, presents the pivotal idea of this book, the Information-Rich Engineering Design (I-RED) model. This model synthesizes concepts from the first three chapters to create a generic model of the elemental activities in engineering design and the corresponding information-seeking and -creating activities.

Part II, Chapters 5 through 14, provides specific practical advice and tools on how students can be guided in learning to manage and integrate information based on each phase of a design project, from conception to realization, based on the elements in the I-RED model. This includes addressing ethical considerations (Chapter 5) and team and knowledge management decisions (Chapter 6), problem scoping through eliciting user feedback (Chapter 7), gathering background information about the project (Chapter 8), and investigating professional best practices (Chapter 9). It also includes investigating prior art (Chapter 10), evaluating the quality of information and incorporating it to making evidence-based design decisions (Chapter 11), actually searching out materials and components to embody the design concept (Chapter 12), and organizing and documenting evidence so that a convincing argument can be made to support the design concept (Chapter 13). Finally, in order for students (and their organization) to benefit most fully from the design experience, they

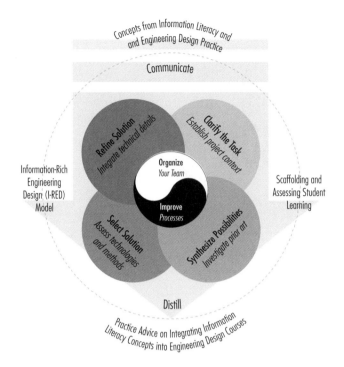

Concepts from Information Literacy and
and Engineering Design Practice

Communicate

Refine Solution
Integrate technical details

Clarify the Task
Establish project context

Information-Rich
Engineering
Design (I-RED)
Model

Organize
Your Team

Improve
Processes

Scaffolding and
Assessing Student
Learning

Select Solution
Assess technologies
and methods

Synthesize Possibilities
Investigate prior art

Distill

Practice Advice on Integrating Information
Literacy Concepts into Engineering Design Courses

FIGURE I.1   Roadmap for this handbook.

must reflect on the process and identify lessons learned and opportunities to improve processes (Chapter 14). This material is broken out by stage of the design process most relevant for the information activities to enable engineering educators and engineering librarians to support students as they learn to use information effectively as an integral part of doing design. Part III, Chapter 15, offers guidance on how to prepare students to incorporate information into engineering-related decision-making activities as a precursor to full-on informed design projects and how to assess student learning outcomes.

A particular feature of this handbook is that each chapter begins with a list of expected learning outcomes. This approach reflects good pedagogical practice and is intended to explicitly orient readers at the outset to the things they should be able to do after actively engaging with the content of each chapter. The best way for

readers to accomplish the learning objectives is to go beyond just reading the material and to experiment with it in their own educational practice and to use the suggested reading lists to explore the topics covered more broadly. Figure I.1 provides a conceptual roadmap for this handbook.

Throughout this book the term *design* is used intentionally as a verb (the action of designing) rather than as a noun (the outcome of that action). This was done to emphasize the fact that design is an activity, a process, rather than a product. This distinction is made not only to avoid confusion but also to highlight the creative and imaginative act of design. This focus on the act of design is reflected in the choice of verb-noun chapter titles in Parts II and III.

The contents of this handbook can be used to embed information literacy in a standalone design course such as an introduction to engineering project course in the first-year or a capstone design experience. Equally, the tools and techniques presented can be deployed throughout a year-on-year design sequence, from first year to final year. This latter application enables increasingly sophisticated knowledge and skills about the use of information in design to be developed and reinforced over an extended period.

The types of design information referred to are not limited to the obvious sources such as materials selection data, commercial off-the-shelf components and products, patents, and other archived text-based materials that are usually associated with design work. On the contrary, this book strives to include the broadest possible range of types of design information which are gathered in diverse ways and stored in many forms of media. For example, it includes information gathered from the clients and users through interviews and observation and from the literature on local demographics, sociopolitical factors, culture, and geography. Such information might be in the form of field notes, sketches, photographs, videos, maps, statistical data, and so forth.

Design information is also taken as being embedded in physical objects, such as existing artifacts of all types, and physical and virtual prototypes made during the design process to test ideas, as well as resultant components, products, or systems. Similarly, software used in, or resulting from, a design project contains design information. This includes the database of information from the design project itself.

A central tenet of this book is that design is a learning activity whereby existing information is consumed and new information is created. In the process, new knowledge is constructed by each of the parties involved—the client, users, and other stakeholders, members of the design team, and people involved in the final realization of the design solution, as well as others who come in contact with the design solution throughout its life cycle.

Throughout this handbook we have endeavored to keep the tone informal and readable and, ultimately, practical. If we have succeeded, readers should be able to incorporate new activities into their courses that encourage students to take a more informed approach to their design projects, which will then lead to more grounded, practical, and higher quality solutions.

In order to keep this book current, we are maintaining an online site (http://guides.lib.purdue.edu/ired) with materials and suggestions for using the I-RED model.

PART I

Information-Rich Engineering Design

CHAPTER 1

MULTIPLE PERSPECTIVES ON ENGINEERING DESIGN

David Radcliffe, Purdue University

Learning Objectives

So that you can provide students with a robust and holistic appreciation for the engineering design process, upon reading this chapter you should be able to

- Describe the act of engineering design from multiple perspectives: as a process, as critical thinking, as learning, and as a lived experience
- Articulate major factors that lead to successful engineering design

INTRODUCTION

Design is a defining characteristic of engineering. Theodore von Kármán, the Hungarian-born physicist and engineer, is reputed to have said, "Scientists study the world as it is; engineers create the world that never has been." Engineers share this creative endeavor with many other design professionals, ranging from fashion and graphic designers to architectural and industrial designers. While engineers and engineering educators often define engineers as *problem solvers*, this epithet fails to adequately capture the full richness of what it is to engineer (Holt et al., 1985).

Engineering design is a recursive activity that results in artifacts—physical or virtual. These may be new to the world or simply variants on already existing things. Design involves both the use of existing information and knowledge and the generation of new information and knowledge. For engineers, designing is both a creative and a disciplined process. Design requires leaps of the imagination, intuitive insight, the synthesis of different ideas, and empathy with people who come in contact with any new product, system or process that is designed. Yet it also demands careful attention to detail, knowledge of scientific principles, the ability to model complex systems, judgment, a good understanding of how things can be made, and the ability to work under severe time constraints and with incomplete information and limited resources.

For engineers, design is an interdisciplinary undertaking. The variety of disciplines involved extend beyond branches of engineering and can include people with backgrounds in the liberal arts and humanities, as well as other technical disciplines from the biological and the physical sciences.

Design is learned by doing and reflecting. It is not formulaic; it is an art rather than a science.

In the literature the term *design* is used to describe both the act of designing and the resulting artifact (product, system, or service) or the information that fully describes it. To avoid possible confusion, in this handbook we use *design* to describe the action (as a verb), not the outcome (as a noun) (Ullman, 2009).

WAYS TO THINK AND TALK ABOUT ENGINEERING DESIGN

There is no universally agreed upon way to describe the engineering design process. Textbooks on engineering design typically include some form of model that sets out the process as a series of steps or stages with feedback loops and iteration (Dym & Little, 2004). Some of these models attempt to describe the various stages in a general sense, while others are more prescriptive and give considerable detail about the various activities to be undertaken and in what order (Cross, 2008).

Descriptive and Prescriptive Models of Engineering Design

Both descriptive and prescriptive models of engineering design embody a sense of flow or progression, typically shown as a series of steps or stages from top to bottom of the diagram depicting the model. They usually begin with a process of need finding and/or problem analysis and clarification, move to the generation of concepts and then the selection of a preferred concept, followed by the fleshing out or embodiment of this preferred concept into a preliminary solution which in turn is developed into a detailed solution. At each sequential stage, more is known about the artifact being designed; it is much more defined, meaning we have more information about it. This movement or progression through the stages is accomplished by feedback and iteration, as new information causes earlier information to be updated with consequential development of the ideas and information defining the artifact.

Figure 1.1 depicts a typical descriptive model of the engineering design process (French, 1971). The circles represent the information known before and after every stage. This may be in a wide variety of formats: text, drawings, sketches, photographs, moving images, physical models, prototypes or mock-ups, physical artifacts, or computer models and/or simulations. The rectangles represent actions or process steps, each of which have information as inputs and in turn result in new information, often in quite different formats. The lines and arrows indicate the flow of information including feedback to previous process steps, indicating the iterative or recursive nature of design.

Descriptive models present a general overview of a design process without going into many details. The purpose is to give a sense of the major milestones or stages. This type of model is used in most engineering design textbooks in the North America, the United Kingdom, Australia, and other countries whose education is in the Anglo-Saxon tradition. In contrast, the tradition in part of Europe is to teach prescriptive design methodologies. While this tradition goes back nearly a century it is only in the past 20 years that prescriptive models have become widely discussed in the English-speaking world.

Emblematic of this prescriptive approach is the classic text by Pahl and Beitz (1996). As illustrated in Figure 1.2, the broad stages of design—for example, clarify the task or

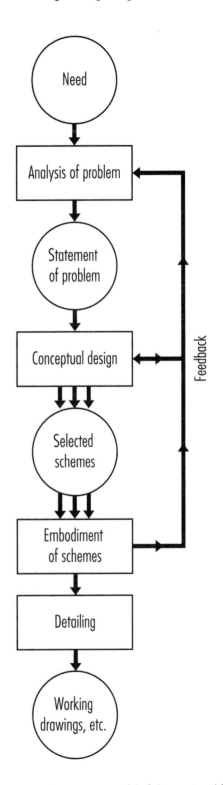

FIGURE 1.1   Descriptive model of design. (Modified from French, 1971.)

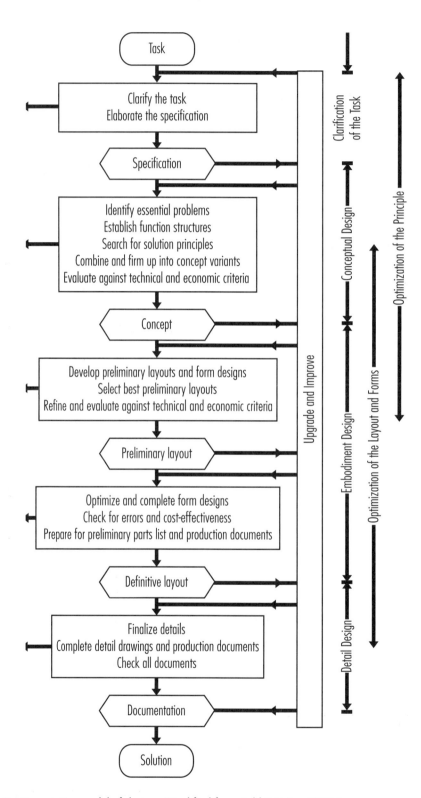

FIGURE 1.2   Prescriptive model of design. (Modified from Pahl & Beitz, 1996.)

conceptual design—are indicated on the right-hand side of the model. Each stage is broken down into a set of discrete tasks as listed in the rectangular boxes. Each stage takes in information from the preceding one, creates additional information, and in turn provides this to the subsequent stage. These sets of information are shown in the boxes with the pointed ends. The iteration is indicated by the upgrade and improve band and the horizontal arrowed lines. Information flows are explicitly indicated by the dotted line on the left-hand side of the diagram.

While this model looks superficially similar to a descriptive model, there is much more detail, including the step-by-step list of design tasks. Moreover, this diagram is only a high-level summary. Pahl and Beitz (1996) and similar textbooks devote whole chapters to each stage and go into considerable detail in setting out how each task should be carried out and the sorts of design techniques that are most appropriate to accomplish each task. For instance, the conceptual design phase has five steps in this high-level model: (1) identify essential problems; (2) establish function structures; (3) search for solution principles; (4) combine and firm up the concept variants; and (5) evaluate against technical and economic criteria. However, in the detailed model of conceptual design, each of these expands to several subtasks. Further, the level of detail and specificity around topics like conceptual design, solution principles, and the principles of embodiment design is much higher than that found in a traditional engineering design textbook used in North America, where there is much more emphasis on component design (machine elements in mechanical design). That said, there has been a trend in recent years to incorporate more system-level and systematic design ideas in many engineering design textbooks.

Design as a Learning Activity

An alternative way to think about the engineering design process is as a learning activity. Learning is effectively a change in our state of knowledge or understanding. As previously mentioned, design is inherently an iterative process during which information is consumed and new information and knowledge about the task and/or the prospective product, system, or service being designed is acquired by the design team. As they progress through a project, design team members continuously learn more and more. In its most fundamental form this comes down to the team's having ideas which are tested or validated by an appropriate means. Often testing of their ideas produces outcomes that were not as originally anticipated. As the team interprets and reflects upon the results of these tests, such dissonance causes them to learn something new about the project. This is illustrated in Figure 1.3.

This idea-test cycle is repeated at every stage of a design project from clarifying the task all the way through to documenting and communicating the final, complete description of the product, system, or service created. At each of these project stages the sources of ideas and the means of arriving at them may vary greatly. Figure 1.3 indicates only a few of the possible idea generation strategies.

Having neat ideas is not sufficient; they must be put to the test to see if they perform as imagined. This requires the team to act on the ideas in a way that will subject the ideas to scrutiny in a way that will assess their veracity. As with idea generation, testing takes place in varying degrees throughout the design project. This can be something as simple as a thought experiment or a simple prototype made from bits and pieces at hand all the way up to, say, the flight-testing of a new concept of aircraft.

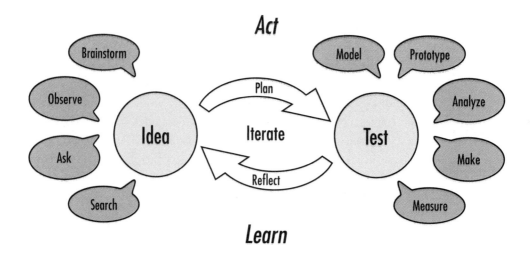

FIGURE 1.3 Idea-test-learn model of design.

Types of testing can include modeling and analysis, simulations, physical mock-ups, working prototypes of subsystems or assemblies, or early prototypes. The *design thinking* movement espouses that the prototyping of ideas be done early and often (Brown, 2009). This accelerates the learning process by going through a large number of idea-test-learn cycles in a short period of time.

Similarly, it is not sufficient to merely test an idea or a system; the findings have to be reflected upon critically so as to extract the deep and lasting lessons to be learned. This is not as easy as it sounds. It takes a disciplined approach and an inquiring, sometimes skeptical mind. The learnings need be captured, kept, communicated, and acted upon as appropriate throughout the remainder of the project. Some of this knowledge may be vital across the whole life cycle of the artifact being designed.

Design as Critical Thinking

Engineering design is not an exact science that has single, absolute, immutable answers. Rather it is a situated and contingent activity. Engi-

neers have to develop the confidence and the courage to make professional judgments on the basis of evidence and argument. They have to be able to make tough calls that can literally have life and death consequences and be prepared to live with those consequences. This requires critical thinking of the first order.

Even if a prescribed methodology is adopted, the design process requires engineers to make simplifying assumptions so that the creative work can proceed. They must step from the *physical world*, where the laws of nature apply, to the *model world*, where it is not possible to simulate every aspect of the behavior of even an ideal system. Subsequently, engineers make critical decisions on the basis of these assumptions and incomplete information. The availability of design information is limited by many factors, including available time, finite human resources, gaps in knowledge (especially in cutting edge projects), ready access to timely and up-to-date information, and the ability to adequately communicate what is known. This cycle is depicted in Figure 1.4.

Design as critical thinking depends upon the team's ability to model the prospective per-

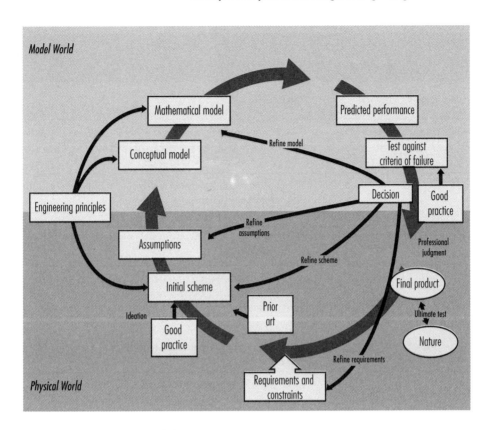

FIGURE 1.4   Design assumptions and decisions.

formance of proposed concepts and systems using prototyping and simulation. While the level of sophistication and completeness and hence veracity of such modeling and simulation continues to improve, models are only ever an approximation to reality. This is due to a combination of our ability to fully describe how complex technical, let alone sociotechnical, systems behave and the uncertainty in the values of the properties of the components. Professional judgment is required to both create models and to interpret their outputs. So while many of the tools and techniques that engineers use when designing are powerful and precise and rely on scientific knowledge, the overall design process does not have these characteristics. The engineering design process does not have the predictive certainty of science.

Design as Lived Experience

Engineering design is a social activity (Brereton, Cannon, Mabogunje, & Liefer, 1997)—a deeply human activity (Petroski, 1982). While it may be concerned with technological artifacts and knowledge, it is carried out by people, typically from diverse disciplines, working in teams. A number of researchers have studied the human act of designing in fields including engineering (Bucciarelli, 1996) and architecture (Cuff, 1992), complete with the frailties and ambiguity inherent in language and human discourse.

A recent study of designers (Daly, Adams, & Bodner, 2012) working in diverse fields from engineering to instructional design and fashion design used phenomenography

TABLE 1.1 *The Variety of Ways That Design Is Experienced*

Design was esperienced as . . .	Design is . . .
Evidence-based decision making	Finding and creating alternatives, then choosing among them through evidence-based decisions that lead to determining the best solution for a specific problem.
Organized translation	Organized translation from an idea to a plan, product, or process that works in a given situation.
Personal synthesis	Personal synthesis of aspects of previous experiences, similar tasks, technical knowledge, and/or others' contributions to achieve a goal.
Intentional progression	Dynamic intentional progression toward something that can be developed and built upon in the future within a context larger than the immediate task.
Directed creative exploration	Directed creative exploration to develop an outcome with value for others, guided and adapted by discoveries made during exploration.
Freedom	Freedom to create any of an endless number of possible outcomes that have never existed with meaning for others and/or oneself within flexible and fluid boundaries.

Modified from Daly, Yilmaz, Christian, Seifert, & Gonzalez, 2012.

to discover the variety of ways in which designers experience design. The findings are summarized in Table 1.1. The respondents experienced design in one of six broad ways, each characterized by a word or phrase (e.g., evidence-based decision making). The researchers describe each of these six different ways of experiencing design in terms of a short description expressed as *design is . . . .* From the top to the bottom of Table 1.1, there is a progression in the way that design is experienced: from a bounded, procedural experience toward a more unbounded, emergent, learning, and meaning-making experience.

This study suggests that design can be experienced as a relatively defined process of the type depicted in descriptive and prescriptive models of the design (i.e., *evidence-based decision making* or *organized translation*). Equally it can be experienced as a much more personal and nuanced progression of discovery (i.e., *personal synthesis* and *intentional progression*). This is not captured in typical models of design. The final two types of experience are values based and much more about finding creative expression, or empowerment, in a large solution space (i.e., *directed creative exploration* and *freedom*). These different ways of experiencing design impact the types of information sought and generated during a project and often the ways in which this information is captured and communicated.

SUCCESS FACTORS IN ENGINEERING DESIGN PROJECTS

Engineers design in teams in the context of a project. The Project Management Body of Knowledge (PMBOK) (Project Management Institute, 2000, p. 4) defines a project as "a temporary endeavour undertaken to create a unique product or service" or as "an endeavour in which human, (or machines), material, and financial resources are organised in a novel way, to undertake a unique scope of work, of a given specification, within constraints of cost and time so as to deliver beneficial change defined by quantitative and qualitative objectives." The implications of this are that the information needed for a given design project might have to be assembled specifically for the unique circumstances of that project or perhaps repurposed and reconfigured from resources used on similar but different past projects.

Why Engineering (Design) Projects Succeed or Fail

While all engineering projects aim to be successful, the irony is that design failures provide valuable lessons that can underpin future success (Petroski, 1982). Failure of an engineering project, including design projects, can be technical, economic, environmental, or sociocultural. Box 1.1 contains a list of seven frequently occurring reasons for project failure (Eisner, 1997). The first six all depend to a greater or lesser degree on some aspect of how information is discovered, accessed, interpreted, communicated, used, modified, created, captured, curated, and managed.

Based on the analysis of many engineering design projects that resulted in artifacts that

> **BOX 1.1**
> **Why Engineering Projects Fail**
> 1. Inadequate articulation of requirements
> 2. Poor planning
> 3. Inadequate technical skills and continuity
> 4. Lack of teamwork
> 5. Poor communication and coordination
> 6. Insufficient monitoring of progress
> 7. Inferior corporate support
>
> Data from Eisner, 1997.

failed, Hales and Gooch (2004) identified ten strategies (see Box 1.2) that can help engineering designers avoid failures. Attending adequately to any of these implies a sophisticated level of information literacy, in the broadest sense, including an appreciation of the cultural or linguistic assumptions behind information and how it is represented, especially when working in a global context. These success strategies assume the members of the design team appreciate the social and cultural mores and the

> **BOX 1.2**
> **Strategies for Design Success**
> 1. Define the real problem or need
> 2. Work as a team
> 3. Use the right tools
> 4. Communicate effectively
> 5. Get the concept right
> 6. Keep it simple
> 7. Make functions clear
> 8. Make safety inherent
> 9. Select appropriate materials and parts
> 10. Ensure that the details are correct
>
> Data from Hales & Gooch, 2004.

aesthetic sensibilities of diverse user communities. Existing artifacts and depictions of their use are therefore a vital source of information for designers as these objects embed critical social and cultural knowledge. Without this information it is difficult to identify the real problem and a complete set of requirements, communicate effectively, make the functions clear, select appropriate materials, and so forth.

Managing Expectations

Success in design is ultimately about managing expectations. There must be convergence between the perceived needs and the emergent solution, as experienced by multiple stakeholders with differing perspectives. The real need is never fully known at the outset, and perceptions of the need can change over time. Success involves arriving at a mutually agreeable destination rather than being on a predictable journey from A to a B, where B is defined precisely at the outset. This does not imply that design is a random exploration without a target. The idea of managing, as much as meeting, expectations recognizes the contingent nature of design and the reality that the target will change during the course of any nontrivial project as new information emerges or is discovered.

The *PMBOK* (Project Management Institute, 2000) defines project management as the application of knowledge, skills, tools, and techniques to project activities in order to meet or exceed stakeholders' needs and expectations from a project. Meeting the needs of the stakeholders implies that the design team knows who all the stakeholders are in a given project, not simply the client who approaches the designer initially with a brief or a request for proposals, but all those individuals and groups who will come in contact with the product,

system, or service being designed throughout its life cycle—from inception to decommissioning and recycling or reusing the artifact or its elements after its operational life. Thus a designer needs to identify all the potential stakeholders and know enough about them so as to be able to determine their possible needs and expectations. These needs not only are technical in nature but also could draw on cultural, historical, social, geographical, economic, and other nontechnical types of knowledge.

Information literacy is a critical skill in resolving the following set of questions related to managing expectations. What is the scope of the project (what aspects are to be included)? What has been done previously to tackle this need? Are there analogous circumstances we can learn from? What are the roles and responsibilities of the team members? What has to be communicated to whom, and when and how should communication take place, to capture and preserve vital information? How can we create sharable models and other representations of the emergent artifact that are readily accessible for different participating disciplines and stakeholders? What information is there that can help the team to develop into an effective group that sustains high levels of performance?

Dealing with Uncertainty

Design projects of any substance are complex in the sense that they exhibit emergent properties. At the commencement of any project it is impossible to have complete knowledge of everything that might happen nor every piece of information that might be needed. During a design project it is not possible to predict completely nor with perfect precision how the product, system, or process being designed or its component parts or assemblies will behave under all possible circumstances. Accordingly, engineers must be comfortable with ambiguity and be able to handle uncertainty. These related abilities are bound up in the concept of risk and risk management. The *PMBOK* defines risk management as the "processes concerned with identifying, analyzing, and responding to risk [throughout the project life cycle]. It includes maximizing the results of positive events and minimizing the consequences of adverse events" (Project Management Institute, 2000, p. 127). Risk is a combination of the frequency (or probability) of occurrence *and* the consequences of a specified (hazardous) event.

Examples of the types of risks that frequently impede the success of engineering design projects listed in Box 1.3.

BOX 1.3

Engineering Design Risks

1. Insufficient or inappropriate personnel or project plan
2. Requirements not adequately identified or defined
3. Noncompliance of system to requirements
4. Program scope increases due to requirements creep
5. Using unproven technology
6. Poor knowledge management or poor quality systems
7. Delays in procurement of materials or parts
8. Materials do not meet the specification
9. Insufficient infrastructure for integration schedule
10. Technical performance not supportable in field
11. Reliability inadequate or issues with logistics
12. System not maintainable to end of program or life cycle

Each of these risks has a critical information dimension. Reducing the uncertainty and hence managing these risks is highly dependent upon having the most complete and accurate information available at the time it is really needed, tracking key information and its interdependence upon design decisions, being able to locate the right information quickly and easily when required, keeping information up to date, and preserving the integrity of information over the life cycle of a product, system, or service.

Grasping Opportunities

The counterpoint to risk is opportunity. From uncertainty there may arise opportunities to do things a different way or to take the project in a different, more fruitful direction. Grasping the upside of uncertainty can be just as important to the success of a design project as managing the potential downside of risks. Indeed, many national and international standards on risk management actually cover both risk and opportunity management. Unfortunately, the overwhelming bulk of the material in such standards focuses on risk, which is a reflection of the designer's imperative to avoid being responsible for a foreseeable fault or problem in a project outcome.

Strategies for making the most of potential opportunities in design include the following: using modern value engineering or value management techniques to continuously seek better ways to do things; negotiating changes to the project scope to enable alternative solutions to apply (e.g., solutions that that reduce the life cycle cost, better meet requirements, or meet implicit client/stakeholder needs); freeing up project constraints to enable alternative approaches/ solutions; and broadening the search of solutions to similar problems to reveal new technologies or approaches that open up out-of-sector solutions.

Measures of Success

A simple way to consider the success of a design project is to use the three generic criteria espoused by the internationally renowned new product development firm IDEO (Brown, 2009): user desirability, technical feasibility, and business viability. A successful product, system, or service must meet the actual needs of the prime user and more generally consider all of the people who will encounter it during its life cycle—from conception to recycling. That is, the approach to design should be human centered (Donald, 1988). Second, products, systems, or services can only be successful if the underlying technologies are sufficiently capable and robust enough to ensure safe, reliable operation. Innovative design concepts can be ahead of their time in the sense that the most appropriate technology does not yet exist to enable the idea to be effectively realized. Finally, a product, system, or service must also be viable in terms of its whole of life cost—not just the purchase price in relation to the production cost. Further, there must be a viable business model in place. Business success can be measured in pure dollar terms or other ways as appropriate. To be successful, the design solution must deliver sustainable value when viewed from all three of these perspectives, not just one or two of them.

Safety, Clarity, and Simplicity

One design strategy that can help to achieve this sustained value is to ensure that the chosen concept and the way it is embodied meets the following three basic criteria: safety, clarity, and simplicity (Pahl & Beitz, 1996).

Safety. The concept and its form should be inherently safe. It should not be necessary to design in safety features as an afterthought during detailed design in order to overcome problems that could have been avoided in the earlier stages of the project.

Clarity. The operation of the product, system, or service should be obvious to the users and clear for them to easily understand, even intuitive. Clarity in the form and function is also critical for people other than users (e.g., maintenance personnel) who must work with the product, system, or service at any point during its life cycle.

Simplicity. In essence, keeping things simple often results in artifacts that are easier and less expensive to manufacture, as well as easier to maintain. This is also known as the KISS principle: Keep it Simple for Success. Apple products are an excellent contemporary example of simplicity deployed as the guiding design philosophy (Segall, 2012).

Engineers have been known to design things that are unnecessarily complicated or have too many bells and whistles when a much more straightforward solution would have sufficed (Thomke & Reinersten, 2012). Mark Twain is reputed to have apologized for sending his friend a long letter as he did not have time to write a shorter one. Similarly, it is much more difficult to create a product, system, or service that is inherently safe, clear to understand, and simple to make or use than it is to create an overly engineered artifact.

The last word in design success comes from physicist and Nobel laureate Richard Feynman. In a famous minority appendix in the Rogers Commission Report on the explosion of the space shuttle *Challenger*, Feynman (1986) made an important and sobering distinction between reliance upon authentic information rather than mere rhetoric in making critical design or operational decisions: "For a successful technology, reality must take precedence over public relations, for nature cannot be fooled" ("Conclusions," para. 5).

IMPLICATIONS FOR STUDENT DESIGN PROJECTS

In learning to design, engineering students expect some guidance on what to do, when to do it, how best to do it, and so forth. It is clear that while engineering design is often represented as a multistage process with iterations, the reality and the experience of a real design project is much more human, contingent, and complex. While the teaching of specific design techniques (e.g., brainstorming) and analysis tools (e.g., computer simulation) might be amenable to instructional techniques, the overall process of conducting a design project is much more elusive and therefore almost impossible to teach. Those from the European tradition of design education constructed around prescriptive design models would argue that the overall process of engineering design can be taught.

Many experienced design educators have found that teaching design is more about coaching individuals and student teams through a series of scaffolded learning experiences preferably based on authentic design tasks. This is easiest to achieve if there are regular design experiences spread periodically across the curriculum (e.g., one every semester) and if these are centered on increasingly challenging tasks—challenging either in the scope or in the scale of the project. This approach also affords the opportunity to develop and integrate a breadth and depth of corresponding information literacy skills over a multiyear period. Of course, this professional growth and development continues beyond the completion of college and spans a career.

The methods and tools available in engineering practice and how and when these are used are not the same as those for a typical student engineering design team. Most students would be classified as novice designers with limited experience. Furthermore, the range and diversity of design and other professional experience in a student team is narrow, even if the students are enrolled in quite different majors. For university-based projects, typically there is little in the way of "corporate memory," such as comprehensive documentation of past projects, lists of lessons learned, or even cogent advice on the best ways for approaching and managing projects. While some design researchers have developed and assessed the use of electronic repositories and knowledge exchanges with student design teams, this is the exception rather than the norm. In contrast, teams in industry have access to very sophisticated company- or even industry-wide Web-based collaboration tools that enable sub-teams of specialists from around the globe to participate and which have vast stores of product information data and test data. These differences between the working environment of student design teams as compared with that of engineering practitioners poses some interesting challenges and indeed opportunities for how we develop effective information literacy interventions in engineering schools and associated technologies to foster and support good information practices that carry beyond the classroom.

SUMMARY

There are many approaches to experiencing engineering design, including process-oriented, human-oriented, and learning-oriented. However, whichever way engineering design is taught, it is intrinsically a complex activity and, while structured, is ultimately creative as well.

It thus requires the integration of many information inputs, synthesis, and analysis, which results in the construction of something that has not existed before. In order to ensure the best chance of success in completing a project to the expectations of the clients, information needs to be gathered, organized, and applied appropriately, ethically, and efficiently. Like other professional skills, information management skills need to be addressed in the engineering curriculum to ensure that students can create rich solutions to the design challenges they will face in their professional careers.

REFERENCES

Brereton, M. F., Cannon, D. M., Mabogunje, A., & Liefer, L. J. (1997). Collaboration in design teams: Mediating design progress through social interaction. In K. Dorst (Ed.), *Analyzing design activity*. New York: Wiley.

Brown, T. (2009). *Change by design: How design thinking transforms organizations and inspires innovation*. New York: HarperCollins.

Bucciarelli, L. L. (1996). *Designing engineers*. Cambridge, MA: MIT Press.

Cross, N. (2008). *Engineering design methods* (4th ed.). New York: John Wiley & Sons.

Cuff, D. (1992). *Architecture: The story of practice*. Cambridge, MA: MIT Press.

Daly, S. R., Adams, R. S., & Bodner, G. M. (2012). What does it mean to design? A qualitative investigation of design professionals' experiences. *Journal of Engineering Education, 101*(2), 187–219. http://dx.doi.org/10.1002/j.2168-9830.2012.tb00048.x

Daly, S. R., Yilmaz, S., Christian, J. L., Seifert, C. M., & Gonzalez, R. (2012). Design heuristics in engineering concept generation. *Journal of Engineering Education, 101*(4), 601–629. http://dx.doi.org/10.1002/j.2168-9830.2012.tb00048.x

Donald, N. (1988). *The design of everyday things*. New York: Basic Books.

Dym, C. L., & Little, P. (2004). *Engineering design: A project-based introduction* (2nd ed.). New York: John Wiley & Sons.

Eisner, H. (1997). *Essentials of project and systems engineering management*. Hoboken, NJ: John Wiley & Sons.

Feynman R. P. (1986). *Report of the Presidential Commission on the Space Shuttle Challenger Accident: Appendix F—Personal observations on the reliability of the shuttle*. Retrieved from http://science.ksc.nasa.gov/shuttle/missions/51-l/docs/rogers-commission/Appendix-F.txt

French, M. J. (1971). *Engineering design: The conceptual stage*. London: Heinemann Educational Press.

Hales, C., & Gooch, S. D. (2004). *Managing engineering design* (2nd ed.). London: Springer-Verlag. http://dx.doi.org/10.1007/978-0-85729-394-7

Holt, J. E., Radcliffe, D. F., & Schoorl, D. (1985). Design or problem solving—A critical choice for the engineering profession, *Design Studies, 6*(2), 107–110. http://dx.doi.org/10.1016/0142-694X(85)90020-1

Pahl, G., & Beitz, W. (1996). *Engineering design: A systematic approach* (K. Wallace, L. Blessing, & F. Bauert, Trans., 2nd ed.). K. Wallace (Ed.). London: Springer-Verlag.

Petroski, H. (1982). *To engineer is human: The role of failure in successful design*. New York: Wiley.

Project Management Institute. (2000). *A guide to the Project Management Body of Knowledge—PMBOK® Guide: 2000 Edition*. Newtown Square, PA: Project Management Institute.

Segall, K. (2012). *Insanely simple: The obsession that drives Apple's success*. New York: Portfolio Hardcover.

Thomke, S., & Reinertsen, D. (2012). Six myths of product development, *Harvard Business Review, 90*(5), 84–94.

Ullman, D. G. (2009). *The mechanical design process* (4th ed.). New York: McGraw-Hill.

CHAPTER **2**

INFORMATION LITERACY AND LIFELONG LEARNING

Michael Fosmire, Purdue University

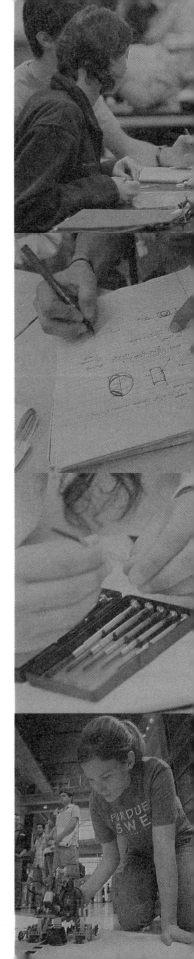

Learning Objectives

So that you can guide students to appreciate the role of information literacy in learning, upon reading this chapter you should be able to

- Articulate four fundamental outcomes of information literacy
- Describe how information literacy relates to critical thinking, problem-solving skills, and lifelong learning
- Understand how the Information Search Process (ISP) model describes the information gathering processes used by students

THE NEED FOR INFORMATION LITERACY

The previous chapter identified different conceptual approaches to engineering design and some of the factors that can improve successful design outcomes. One of the recurring themes is the need for strong information management skills, what librarians commonly refer to as *information literacy*. With the explosion of information technology capabilities, the availability of vast amounts of content on a user's desktop, and the concept of the new generation of "digital natives," who are supposed to navigate these resources effortlessly (Prensky, 2001), instructors can be lulled into believing that they don't need to guide students in locating information resources, let alone understanding and extracting information to be used in their projects.

However, instructors keep complaining that students can't write papers, use poor sources, and have trouble documenting those sources (often resulting in plagiarism, made easier to commit by cutting and pasting text from the Web, and to detect by cutting and pasting suspicious passages into a search engine). With all the information purportedly available, our future engineers still have challenges incorporating information effectively into a report, project, or presentation and solving complex problems.

In the professional sphere, engineers struggle to manage and apply information effectively to solve design problems, leading to delays in product development, overreliance on rules of thumb and prior knowledge that reduces innovation and application of cutting edge technologies, and reinvention/reconstruction of knowledge, all of which lead to reduced profits and competitiveness for the company. Timeliness, accuracy, accessibility, cost, and relevance,

in addition to the core content itself, can be barriers to appropriate integration of information by engineers (Court, Culley, & McMahon, 1997; see also Chapter 3).

There are several definitions and models of information literacy, such as the United Kingdom's Society of College, National, and University Libraries (SCONUL) Seven Pillars of Information Literacy: identify, scope, plan, gather, evaluate, manage, and present (SCONUL, 2011) and the Big6 approach geared toward K-12 students: task definition, information-seeking strategies, location and access, use of information, synthesis, and evaluation (Eisenberg & Berkowitz, 2000). However, the definitions have substantial overlap. For the ease of discussion, in this handbook we will focus on the Association of College and Research Libraries (ACRL) definition widely used by universities in the United States, that information literacy encompasses the ability to "recognize when information is needed and have the ability to locate, evaluate, and use effectively the needed information" (American Library Association, 1989, para. 3). Locate, evaluate, and use effectively each indicate a facet of the information gathering process, and each is essential to the research process.

FACETS OF INFORMATION LITERACY

Recognizing the Need for Information

Of course, without a recognition of the need for information, the search for information never starts. Beyond that, if students cannot articulate what specific information they need, and what information they already possess, they typically resort to ineffectual, often one-word search strategies. We the authors see the same websites crop up on student papers because they are in the first five hits of a Google search on climate change, or electric cars, for example. Trusting Google to do the thinking for them can lead to disastrous results. Rather than seeking out information to confirm or refute theses or fill in gaps in knowledge, many students just try to mix and match their top five sources of information into a report, letting the results determine their research question, rather than their question determine their search for information.

Alternatively, when students first try to scope out a problem, analyze it to determine what they know and what they don't know (including, sometimes, the foundational subject knowledge), they can actually use sources to inform the solution to their problem. They may find general information to get a sense of the big picture before delving into a particular potential solution. With an increased vocabulary, they can use more targeted search terms and use their new knowledge to quickly determine whether a particular source is helpful or even relevant to their problem.

Locating Information

One typically does not think about the ability to locate information as a challenge for students in the Internet age. After all, with several billion pages (certainly more than any one person could possibly hope to look at in their lifetime), the open Web, that is, the part anyone can freely access, would seemingly contain the answer to any question. Digital natives, having grown up with the Internet, are supposed to effortlessly navigate through it. However, more recent findings seem to indicate that students overestimate their information technology abilities and that they have less developed skills than was previously thought (Holliday & Li, 2004). Students rely heavily on the open Web, which is success-

ful for certain kinds of information, such as the weather, stock prices, or even troubleshooting computer problems. As students begin more scholarly and sophisticated inquiries, however, the ability of the open Web to provide the depth of information they need is insufficient.

While many high-quality information sources exist on the open Web, including a large amount of federal and state government information, the bulk of scholarly journals, handbooks, data sources, and books, what we generally think of as traditionally published materials, even if electronic, are behind subscription walls. Indeed, a research library spends several million dollars a year providing access to just these resources. Understanding how and where to find information that is valuable enough to sell, rather than just give away, provides a large conceptual leap for many students.

Locating information requires not only looking in the correct place (the open Web, an index of journals, perhaps a government database or a product spec sheet), but also navigating through that resource to find the specific information needed. Using appropriate search terms and logic, implementing logical search strategies to refine results, understanding how to take advantage of the functionality of different search systems, and capturing and organizing the results all make locating information easier and more effective.

Evaluating Information

Once they have located information resources, students must determine which ones to use and how to use them. They must establish the validity, authority, and relevance of sources rather than taking the information at face value. Students should look for resources with different perspectives, even if just competing products, so that they can critically think about which sources make the most convincing arguments and how those claims can be substantiated or refuted. In general, people remember facts but to a much lesser extent the source of those facts. As a result, a concept can become integrated into one's working knowledge without it ever having been vetted as a reliable piece of information.

Novice information seekers tend to treat any text as reliable, whether from expediency or from a lack of discriminatory skills. Without a well-formulated process for vetting a text—for example, determining the background of the author, whether the author is writing in a field of his or her expertise, or corroboration from other experts in the field—students see every author as having equal standing and may not be able to resolve conflicting claims. Consequently, students will determine that a text that agrees with their prior preference or conveniently fits their thesis is the most reliable. Alternatively, students may consider the competing claims to be a matter of opinion and not seek to determine which side has a more valid argument (King & Kitchener, 1994).

Once they have sufficiently analyzed information from a source, students need to determine whether it matters. Is the information convincing enough that they are willing to change a deeply held belief? Is it important enough to incorporate into their working knowledge? Is it something that they believe in enough to stake a professional or personal relationship on? Without a conscious engagement with the information on a deep level, facts remain facts and are not transformed into knowledge.

Applying and Documenting Information

Once information has been located and deemed credible, it needs to be applied to inform the solution to the original problem. Students must extract the particular information relevant to the problem and then organize, synthesize, document, and communicate that information. Unless something is done with the information, it remains in a state of abstraction—as interesting facts rather than usable knowledge.

Extracting appropriate information from a text first requires students to understand what they are reading. This means that students need to find information that is at an appropriate level for them. First-year students likely will find scholarly texts incomprehensible, so they need to be steered to the kinds of resources written at their level. When asked to explore more advanced concepts, students should be directed to overview articles, technical encyclopedias, or other background sources to obtain context and conceptual foundations from which to build a deeper understanding. Techniques such as note taking and restating or discussing with peers provide opportunities for students to go beyond the passive intake of information and to transform it into an active engagement and synthesis of the content.

In addition to understanding an information source, students also need to use information ethically and appropriately. Contrary to current political discourse, in which increasingly the goal appears to be creating impressive sound bites without regard to accuracy, in the scientific and technical spheres, persuasion, while still important, needs to be grounded in solid fact. Bridges will not remain standing because of pithy quotes or convenient cherry-picking of facts. Rather, tragedies will only be avoided if a bridge is built according to standards and within the limits of the materials and methods employed in its construction.

In order to ethically use information, then, students need to understand what it is they are asserting, whether the information is credible, and under what conditions it is valid. Students might report a particular value for a material property but not indicate at what temperature or pressure, at what atmospheric condition, the property was measured in. In a more trivial example, a student was calculating the cost savings for moving to a more efficient lighting system. She found a website with utility rates and calculated the expenses without realizing the utility rates were for the Northeast rather than the Midwest, which uses completely different fuels (nuclear versus coal) to generate power at substantially different costs.

Another aspect of ethically using information is the appropriate documentation of that information. Students frequently complain about having to cite their sources, without understanding the purpose of doing so (other than avoiding expulsion for plagiarism). By documenting sources of information, readers have the ability to go back to the original source and make their own determination of its credibility. Otherwise, readers can only assume that the student is the one asserting the statement, which could make it seem less credible. In this way, documentation protects the students. It gives them a proxy of expertise they can tap into, so that the reader can dispute those experts, rather than the expertise of the student. However, it does not stop the reader from disputing how information gained from sources was applied by a student, or questioning the student's judgment regarding whether a particular person is in fact an expert.

Appropriate documentation also allows students to go back to the original source material itself, rather than trying to remember where they found a piece of information. Let's say a proposal to build a project has been accepted. A student may, instead of just reporting that it is possible to build a part with a particular set of properties, actually need to know how to build that part. Instead of trying to reconstruct the previous search for that information, the student could just look back at the references to find the details of fabrication.

LEARNING HOW TO LEARN

Tightly connected to information literacy is the notion of lifelong learning. Once out of the academy, and despite the availability of conferences, workshops, advanced degrees, and online course work, the bulk of professional learning takes place individually and informally. The development of self-directed learning skills, then, becomes paramount to the continued success and viability of engineering professionals in the workplace. Knowles (1975) requires that self-directed learners identify their learning need, determine a learning plan to acquire the skills or abilities to meet the need, actually implement the plan, and be able to determine whether they met their learning goals.

The Knowles (1975) model of self-directed learning mirrors that of information literacy, where, for example, Knowles's *learning need* translates as *recognizing the need for information*. Not all self-directed learning requires a search for information, and not all information gathering activities are self-directed, but the core concept of learning something new to address a specific need provides a large degree of overlap in pedagogy.

The National Academies publication *How People Learn* (National Research Council, 2005) presents three main findings, all of which relate to the absorption of information and the creation of new knowledge. The first finding is that students "come to the classroom with preconceptions about how the world works" (p. 2), and if those preconceptions are not engaged and addressed in the presentation of new information, students might, for instance, learn content for a test but still use their core preconceptions outside of the classroom context. This is often referred to as the *transfer problem* in education. In the world of information literacy, this occurs in the evaluation of information and extraction of knowledge from sources. If students treat information only as something they need to finish an assignment, then no real long-term knowledge has been created. Only by reflecting on what the information means, how it relates to their previous knowledge, and whether they should change those beliefs based solely on that knowledge (or subsequent investigation) do students really learn something from the process. In a meta-sense, information literacy itself can be a subject of analysis. Students have preconceived notions about information, whether they believe that all the knowledge of the world is accessible through Google, or whether a one-word search string should enable a search engine to know what they are really looking for. Or, that all websites are created equal and contain reliable information. Without engaging those preconceptions, students may find five scholarly articles to complete an assignment, but for the next class or after graduation, will likely revert to taking the first Google result as the best possible answer to their question.

The second finding in *How People Learn* (National Research Council, 2005) discusses the development of competence. In particular, students need a foundation of factual knowledge, but they also need to "understand [those] facts and ideas in the context of a conceptual framework" (p. 12), and organize that knowledge so it can be used. Fundamentally, this finding addresses the question of how we can turn novices into experts, able to make profound judgments of a situation and ready to enter the professional world. With a solid con-

ceptual foundation, experts can rapidly determine what information is relevant, and thus quickly hone in on the needed information, ignoring superfluous details. Creating an expert mindset is a lengthy process and one that needs to be consciously cultivated, and information processing is central to that development.

Finally, in *How People Learn* the National Research Council (2005) reports that taking a learner-centered, "metacognitive" approach allows students to control their own learning and monitor their progress. If provided the language and tools to question their own understanding and level of competence, students can become expert self-directed learners. The same tools that allow one to determine the validity of a particular source of information— its credibility, authority, and relevance—play an important role in students' developing the metacognitive skills for learning in the classroom and beyond.

A PROCESS MODEL FOR INFORMATION GATHERING

In teaching information literacy and lifelong learning skills, one first needs to understand how students approach the information gathering process. From the previous section, we see that we need to situate learning in a student's experiences. The Information Search Process (ISP) (Kuhlthau, 2004) provides a structure that students can identify with, especially since the ISP includes affective and cognitive characteristics of the information gathering stages and not just a description of tasks undertaken.

The ISP contains six stages: initiation, selection, exploration, formulation, collection, and presentation. Briefly, these stages are defined as follows:

Initiation: when a person first becomes aware of a lack of knowledge or understanding and feelings of uncertainty and apprehension are common.

Selection: when a general area, topic, or problem is identified and initial uncertainty often gives way to a brief sense of optimism and a readiness to begin the search.

Exploration: when inconsistent, incompatible information is encountered and uncertainty, confusion, and doubt frequently increase and people find themselves "in the dip" of confidence.

Formulation: when a focused perspective is formed and uncertainty diminishes as confidence begins to increase.

Collection: when information pertinent to the focused perspective is gathered and uncertainty subsides as interest and involvement deepens.

Presentation: when the search is completed with a new understanding enabling the person to explain his or her learning to others or in some way put the learning to use.

These stages roughly define a research process that starts from problem definition and scoping to topic selection, thesis formation, documentation and, finally, communication. The first three stages are characterized by the search for relevant information, while the last three stages are characterized by the search for pertinent information. While this model may look like it is most relevant for a full-blown research project, even quick lookups of information may require multiple steps in the ISP, especially if the subject area is not very familiar to the student.

Note that the process described here is conceptual and, consequently, does not discuss the particulars of locating, accessing, or evaluating information. Rather, those concepts would be dealt with in the context of the stage of information search in which the student is currently engaged. For example, if students are in the exploration stage of their ISP, they will be looking for more preliminary information such as encyclopedia or review articles to describe the overall topic, while in the collection phase students will likely need to find technical literature or handbooks or similar materials. Instruction targeting the appropriate stage will provide the tools needed at that time for those students.

CRITICAL THINKING, PROBLEM SOLVING, AND INFORMATION

There are several other cognitive theories that impact information literacy skills. The body of knowledge around critical thinking mirrors the evaluation and application concepts of information literacy. The model of reflective judgment described by King and Kitchener (2002) sheds light into the effect of the developmental stage of students on how they interpret information and use it to make decisions. Finally, common fallacies of reasoning lead to inappropriate and potentially unethical use of information. Each of these areas provides insights into the need for information literacy skills, and aspects that need to be considered when teaching those skills.

Critical Thinking

Critical thinking skills are important to every discipline in the academy. Scriven and Paul (as cited in Critical Thinking Foundation, 2011) describe critical thinking as the

intellectually disciplined process of actively and skillfully conceptualizing, applying, analyzing, synthesizing, and/or evaluating information . . . as a guide to belief and action. . . . Critical thinking can be seen as having two components, 1) a set of information and belief generating and processing skills, and 2) the habits based on intellectual commitment, of using those skills to guide behavior. . . . The development of critical thinking skills is a lifelong endeavor. ("Critical Thinking as Defined by the National Council for Excellence in Critical Thinking, 1987," para. 2)

The Association of American Colleges and Universities (2012) has developed a Valid Assessment of Learning in Undergraduate Education (VALUE) rubric for critical thinking as one of the essential learning outcomes for a liberal education that mirrors in many ways the core tenets of information literacy (see Table 2.1).

The correspondence between critical thinking and information literacy skills is quite robust, and many concepts can be easily applied across those domains. As mentioned above, information that isn't applied remains mere inert facts. Similarly, critical thinking isn't complete unless it leads to actions taken in response to the process.

Reflective Judgment

Students come into the university at different stages of cognitive development. For example, many college students are still in the transitional stage between being concrete and formal reasoners, in the Piagetian model. Similarly, King and Kitchener (1994) found that students faced with an open-ended problem exhibit different levels of development in their ability to make judgments about the problem (see Box 2.1). They found that the average student enters the university in a pre-reflective stage and graduates in a quasi-reflective stage. One of the common misperceptions students

> **BOX 2.1**
> **Reflective Judgment Stages**
> ***Pre-reflective*** — Student gains knowledge through firsthand observation or from an authority figure, not through evaluation of evidence. No ambiguity in beliefs.
> ***Quasi-reflective*** — Student acknowledges a level of uncertainty in a claim, usually attributed to missing information. Uses evidence, although not effectively. Believes that judgments are a matter of opinion, rather than the best-reasoned conclusion.
> ***Reflective reasoning*** — Student acknowledges that claims are not certain and makes judgments based on what student evaluates to be the most reasonable conclusions. Willing to reevaluate judgments as new data becomes available.
> Data from King & Kitchener, 2002.

have when using information is that "if it's on the Internet, it must be true." The reflective judgment model defines this behavior as characteristic of pre-reflective thinking. The development of reflective judgment skills goes hand in hand with the development of evaluation and application information literacy skills.

As students seek to extract meaning from information and, further, to act on that information, they need to develop reflective reasoning skills, and instructors need to understand that this is a process that students go through. Students, especially in the first year, typically cannot effectively incorporate information without specific instruction to support those skills (see Jackson, 2008; Pascarella & Terenzini, 2004).

Common Fallacies of Reasoning

When developing critical thinking skills, students need to be aware of common errors of reasoning. When judging the merits of a

TABLE 2.1 *Comparison of AAC&U VALUE Rubric for Critical Thinking and ACRL Information Literacy Competency Standards*

Critical Thinking Facet	Definition	Information Literacy Analog
Explanation of issues	Problem is stated and described comprehensively, delivering all relevant information necessary for full understanding.	Defining information need
Evidence	Information is taken from sources with sufficient interpretation/evaluation to develop a comprehensive analysis or synthesis. Viewpoints of experts are questioned thoroughly.	Locating information efficiently and effectively
Influence of context and assumptions	Thoroughly analyzes own and others' assumptions and carefully evaluates the relevance of context when presenting a position.	Evaluation of information
Student's position	Specific position is imaginative, taking into account the complexities of an issue. Limits of position are acknowledged. Others' points of view are synthesized within a position.	Application of information
Conclusions and related outcomes	Conclusions and related outcomes (consequences and implications) are logical and reflect the student's informed evaluation and ability to place evidence and perspectives discussed in priority order.	Application of information

Data from Association of American Colleges and Universities, 2012.

particular information source, for example, students need to analyze whether the author has made an honest, supported argument, or whether the author has engaged in sloppy or misleading reasoning. Although using rhetorical tricks can be an effective way to influence others in the political arena, because the results of the engineering design process yields artifacts that impact safety, a high standard of information gathering needs to be enforced for students.

A typical example is students collecting product information by using an Internet search engine to find, for example, air conditioners or noise cancellation devices. Commonly students will not systematically attempt to compare products. Instead, they may make their decisions about which device to use based solely on marketing claims, such as customer testimonials or expert endorsements, rather than by evaluating product specifications.

Francis Bacon (1676) developed one of the early categorizations of common fallacies of reasoning. He called them the four idols, which need to be demolished in order to engage in clear and rigorous thinking.

Idols of the tribe. As human beings we have certain physiological and psychological biases in how we observe the world and assign meaning to what we perceive. How we are wired affects how we understand the world.

Idols of the cave. We each live in our own "cave" of individual experience, "where the hight of

Nature is obscured and corrupted" (p. 5). We each have developed our own construction of knowledge, based on what we've read or not, who we've talked to, if we've been in traumatic situations, and so forth.

Idols of the marketplace. Misapprehensions occur in the communication between people in society, as imprecise and "improper imposition of words doth wonderfully mislead and clog the understanding" (p. 5). Ideas can be obscured by the limitations of language to convey those concepts.

Idols of the theater. This refers to the effect of ideologies or systems of thought that are embraced because of "tradition, credulity and neglect" (p. 5), rather than critical examination. Uncritical acceptance of a particular philosophy or scientific model leads to people's arguing about the particulars of the idea and ignoring whether the model is based on solid evidence.

This is not to say that Bacon's idols are without value. For example, the ability for people to make patterns out of data (sometimes erroneously) has survival value, when, for example, the one time in a hundred, it *is* a nefarious person and not an oddly shaped tree trunk you see when walking alone after dark. Questioning everything leaves little time to actually do something. However, when asked to make an important judgment, it is important to understand how well a fact or concept is known and its limits of application.

Since many, especially informal, information sources use faulty logic, we describe in Box 2.2 a few of the most common as examples of what students need to watch out for both in reading and in making their own arguments. Some of these fallacies are intertwined with stages of reflective thinking (for example, appeals to authority), others with sloppy thinking, and sometimes these appeals are used deliberately as rhetorical devices. Rhetoric can be quite influential and effective, but words alone cannot trump physical reality when it comes to developing proficient and ethical engineers.

BOX 2.2
Common Fallacies of Thinking

Ad hominem/appeal to authority—Attacking the person rather than the idea. Either vilifying the character of the person, or, conversely, exalting the person's credentials or morality.

Appeal to common knowledge—Everyone knows something is true; therefore I don't need to justify a particular point.

Appeal to ignorance—If we haven't found something, it must not exist.

False choices—Framing a problem as having only two solutions or two causes, rather than allowing for a variety of options. Usually, one solution is ill-crafted, so the preferred solution is introduced as the one to follow.

Confirmation bias—Discounting occurrences that don't fit a model, and emphasizing occurrences that do.

Proof by example (inappropriate generalization)—If it happened once, it must be true in general.

Repetition—If you say something often enough (or see it enough in print), it is true.

Part to whole—If an item belongs to a group, it has all the properties of other members of the group (not just the group properties).

INFORMATION GOALS FOR ENGINEERING STUDENTS

The ABET (2013) accreditation criteria guides the development of engineering programs. Criterion 3 delineates the student outcomes required of the program (see Box 2.3). Librarians have frequently focused on criterion 3 (i), "a recognition of the need for, and an ability to engage in life-long learning," as the area most aligned with information literacy. However, this potentially relegates information literacy to that which happens after graduation, rather than integrating information literacy directly into the problem solving process for engineers. Riley, Piccinino, Moriarty, and Jones (2009) and Sapp Nelson and Fosmire (2010) both have mapped ABET criteria to ACRL information literacy standards. While their analysis is not repeated here in great detail, it is important to understand that information gathering takes place in all but the most trivial of problem solving situations (i.e., except when working computational textbook problems).

Some of the more saliently overlapping outcomes (ABET, 2013; Riley, Piccinino, Moriarty, & Jones 2009; Sapp Nelson & Fosmire, 2010) include the following:

"An ability to design and conduct experiments" (ABET, 2013, "General Criterion 3. Student Outcomes"). Every experimental design includes a literature review as a hypothesis is being formed and frequently when data has been collected and analyzed.

"An ability to design a system . . . to meet desired needs within realistic constraints" (ABET, 2013, "General Criterion 3. Student Outcomes").

BOX 2.3

General Criterion 3. Student Outcomes

(a) An ability to apply knowledge of mathematics, science, and engineering

(b) An ability to design and conduct experiments, as well as to analyze and interpret data

(c) An ability to design a system, component, or process to meet desired needs within realistic constraints such as economic, environmental, social, political, ethical, health and safety, manufacturability, and sustainability

(d) An ability to function on multidisciplinary teams

(e) An ability to identify, formulate, and solve engineering problems

(f) An understanding of professional and ethical responsibility

(g) An ability to communicate effectively

(h) Ahe broad education necessary to understand the impact of engineering solutions in a global, economic, environmental, and societal context

(i) A recognition of the need for, and an ability to engage in life-long learning

(j) A knowledge of contemporary issues

(k) An ability to use the techniques, skills, and modern engineering tools necessary for engineering practice

From ABET, 2013.

"An ability to identify, formulate, and solve engineering problems" (ABET, 2013, "General Criterion 3. Student Outcomes"). Engineering is a human-centered activity, and consequently information must be gathered from stakeholders to understand a problem fully. Furthermore, when meeting the variety of constraints listed, substantial information needs to be gathered about the particular situation in which the students are working so that they can apply their methodologies appropriately and understand the consequences of their decisions.

"An understanding of professional and ethical responsibility" (ABET, 2013, "General Criterion 3. Student Outcomes"). Information ethics (see Chapter 5) are quite important for engineers. How information is documented, communicated, and utilized all have consequences for ethical behavior.

"Broad education necessary to understand the impact of engineering in a global, economic, environmental, and societal context" (ABET, 2013, "General Criterion 3. Student Outcomes").

"Knowledge of contemporary issues" (ABET, 2013, "General Criterion 3. Student Outcomes"). Similar to (c), the engineer needs to be able to find information to maintain currency in societal issues surrounding engineering.

The "recognition of the need for, and an ability to engage in life-long learning" (ABET, 2013, "General Criterion 3. Student Outcomes").

The preceding discussion provides a template for acquiring lifelong learning skills and abilities. In addition, the recognition of the need for lifelong learning is quite analogous to an internalization of the ISP, starting with recognizing the need for information.

INFORMATION LITERACY AND DESIGN

Engineering design provides an ideal situation for practicing information literacy and lifelong learning skills. A typical design problem is *ill-structured*, a term meaning a complex problem without a well-defined solution. As such, the students will, or should, come into contact with concepts, ideas, and details they are unfamiliar with, and a measure of their success will be in finding appropriate information to apply to those problems. Just because a process wasn't mentioned in a textbook doesn't mean it is not the best solution. Indeed, engineering design problems provide the most authentic situations for students to practice skills they will need after graduation, including gathering information in ways that they will likely encounter in their careers after graduation.

SUMMARY

This chapter has introduced a variety of concepts related to cognition, lifelong learning, and information literacy. Information literacy comprises more than just how to find information—it encompasses the ability to understand the need for information, interpret the information, and appropriately apply and document the information. Perhaps most important, information literacy requires metacognitive skills that allow students to make the most of their learning experiences. In order for a student to develop an informed approach to acquiring new skills and maintaining currency in a field, information literacy needs to be a component of his or her lifelong learning strategy. Design projects, as authentic learning activities, are ideal environments for learning the skills necessary for professional success for engineering students.

REFERENCES

ABET. (2013). *Criteria for accrediting engineering programs 2012–2013*. Baltimore: ABET. Retrieved from http://www.abet.org/DisplayTemplates/DocsHandbook.aspx?id=3143

American Library Association. (1989). *Presidential Committee on Information Literacy: Final report*. Chicago: American Library Association. Retrieved from http://www.ala.org/acrl/publications/whitepapers/presidential

Association of American Colleges and Universities. (2012). *VALUE rubric: Critical thinking*. Washington, DC: Association of American Colleges and Universities. Retrieved from http://www.aacu.org/value/rubrics/index.cfm

Bacon, F. (1676). *The novum organum of Sir Francis Bacon, Baron of Verulam, Viscount St. Albans epitomiz'd, for a clearer understanding of his natural history* (M. D. B. D. Trans.). Viewed in Early English Books Online: http://gateway.proquest.com/openurl?ctx_ver=Z39.88-2003&res_id=xri:eebo&rft_id=xri:eebo:image:106723

Court, A. W., Culley, S. J., & McMahon, C. A. (1997). The influence of information technology in new product development: Observations of an empirical study of the access of engineering design information. *International Journal of Information Management, 17*(5), 359–375.

Critical Thinking Foundation. (2011). *Defining critical thinking*. Retrieved from http://www.criticalthinking.org/pages/defining_critical_thinking/766

Eisenberg, M., & Berkowitz, R. E. (2000). *Teaching information & technology skills: The Big6 in secondary schools*. Worthington, OH: Linworth Publishing.

Holliday, W., & Li, Q. (2004). Understanding the Millennials: Updating our knowledge about students. *Reference Services Review, 32*(4), 356–366. http://dx.doi.org/10.1108/00907320410569707

Jackson, R. (2008). Information literacy and its relationship to cognitive development and reflective judgment. *New Directions in Teaching and Learning, 2008*(114), 47–61. http://dx.doi.org/10.1002/tl.316

King, P. M., & Kitchener, K. S. (1994). *Developing reflective judgment: Understanding and promoting intellectual growth and critical thinking in adolescents and adults*. San Francisco: Jossey-Bass.

King, P. M., & Kitchener, K. S. (2002). The reflective judgment model: Twenty years of research on epistemic cognition. In B. K. Hofer and P. R. Pintrich (Eds.), *Personal epistemology: The psychology of beliefs about knowledge and knowing* (pp. 37–61). Mahway, NJ: Lawrence Erlbaum, Publisher.

Knowles, M. S. (1975). *Self-directed learning: A guide for learners and teachers*. Englewood Cliffs, NJ: Cambridge Adult Education.

Kuhlthau, C. C. (2004). *Seeking meaning: A process approach to library and information services* (2nd ed.). Westport, CT: Libraries Unlimited.

National Research Council. (2005). *How students learn: Science in the classroom*. Washington, DC: The National Academies Press.

Pascarella, E. T., & Terenzini, P. T. (2004). *How college affects students: A third decade of research*. San Francisco, CA: Jossey-Bass.

Prensky, M. (2001). Digital natives, digital immigrants part 1. *On the Horizon, 9*(5), 1–6. http://dx.doi.org/10.1108/10748120110424816.

Riley, D., Piccinino, R., Moriarty, M., & Jones, L. (2009). *Assessing information literacy in engineering: Integrating a college-wide program with ABET-driven assessment*. Paper presented at the 116th ASEE Annual Conference and Exposition, Austin, TX.

Sapp Nelson, M., & Fosmire, M. (2010). Engineering librarian participation in technology curricular redesign: Lifelong learning, information literacy, and ABET Criterion 3. AC 2010-875. In *Proceedings of the ASEE National Conference.* Washington, DC: American Society for Engineering Education.

SCONUL. (2011). *The SCONUL seven pillars of information literacy: Core model for higher education.* Retrieved from http://www.sconul.ac.uk/sites/default/files/documents/coremodel.pdf

CHAPTER 3

WAYS THAT ENGINEERS USE DESIGN INFORMATION

Michael Fosmire, Purdue University

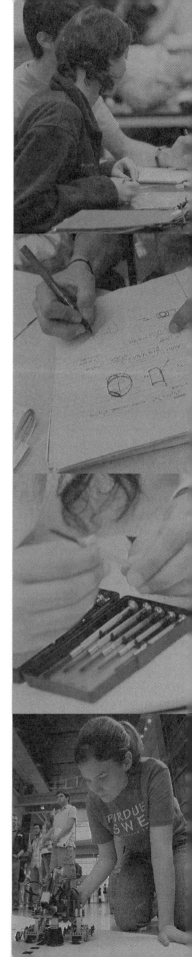

Learning Objectives

So that you can provide students with an understanding of the typical role of information in engineering design, upon reading this chapter you should be able to

- Articulate why engineers gather information and how they utilize it in the design process
- Recognize which information resources are used at different stages of the design process and what information artifacts are produced
- Recognize the main barriers to effective information use by engineers and the role of training in improving their information-seeking behaviors

INTRODUCTION

By understanding the challenges faced by practicing engineers and engineering students in effectively utilizing information to make good design decisions, you will begin to see what gaps need to be filled by instructional interventions. By gaining a deeper appreciation of the competing challenges engineers face, you will see the need to incorporate activities that build information literacy skills in students. Fundamentally, the more familiar and routine information gathering is for students, the more likely they will use those skills in their subsequent work. The observations, models, and opinions in this chapter led us to the development of the Information-Rich Engineering Design (I-RED) model introduced in Chapter 4.

MODELS OF INFORMATION GATHERING

While the library science profession has developed its own models for information gathering, the engineering profession has not neglected the question of the role of information in the design process. Industrial engineering in particular, with its focus on optimizing systems and processes, has provided an extensive body of work looking at particular techniques and information storage and retrieval systems to enhance the outputs of the design process.

Wodehouse and Ion (2010) apply the Data, Information, Knowledge, Wisdom (DIKW) model to the design process (see Table 3.1) to show the transformation of data into knowledge that takes place and the activities that go into that transformation. Briefly, data is the collection of facts and observations available to anyone. The principal activity involved is simply the location of that data. However, value is added by engineers in turning data into information—that is, in organizing it into something usable, making connections between pieces of data, and determining which data are relevant to the problem at hand. Information becomes knowledge when the information is applied to a problem. While information and knowledge are focused on the corporate body or problem under consideration, wisdom is based in the individual, who, by learning in the process of solving the problem, can apply to future problems not only specific content but also the principles and processes used.

Other engineering design models include more concrete analysis of information components. These models incorporate both information inputs and outputs—that is, information gathered from external sources and that produced by the engineers in the course of the design process. Two such models are summarized in Tables 3.2 and 3.3.

Both Ulrich and Eppinger (2011) and Dym and Little (1999) design models recognize that different stages of the design processes call for different information sources, and they explicitly acknowledge that the information process is not only about consuming information but the production of information as well. These models help guide the student through the transformation of data and information into knowledge for the project, with specific activities and processes (i.e., outputs) in the authentic context of engineering design. While neither set of authors spend much time discussing how to access those various kinds of information, Dym and Little (1999) observe that "the *literature review* [emphasis theirs] is so well documented and understood that it might seem un-

TABLE 3.1   *Data, Information, Knowledge, Wisdom (DIKW) in the Design Context*

DIKW Stage	Activity	Design Context	Availability
Data	Locating	Assembling facts	Openly available
Information	Structuring/organizing	Facts are organized and winnowed	Internal
Knowledge	Applying	Information used	Internal
Wisdom	Reflection	Review process; self-assessment	Personal

Data from Wodehouse & Ion, 2010.

necessary for us to comment on it. However, it is worth noting that the relevant literature can be both vast and greatly dependent on the stage or phase of the design" (p. 41). These models provide the structure, through the engineer's lens, for activities that engineers and librarians, working together, can develop to build information gathering skills and, ultimately, an informed design product.

VALUE OF INFORMATION GATHERING

A variety of interview and observation studies indicate that engineers appreciate the role of information gathering in the design process. Mosberg et al.'s (2005) interview of engineers found gathering information to be the fourth most important activity out of 24 components of the design process, below only

TABLE 3.2   *Information Use in the Engineering Design Model of Ulrich and Eppinger*

Design Stage	Information Sourced	Information Generated
Planning	Market data, company reports	Briefing documents, project plan
Concept development	Competitor and related products, previous design schemes	Brainstorming notes, sketches, drawings, rough calculations
System design	Patents, previous design schemes	Sketches, drawings, mock-ups and models, cost evaluation
Detail design	Textbooks, catalogs, suppliers' data	Detailed drawings and design calculations, solid and mathematical models
Testing	Standards, databases	Experimental data, manufacturing drawings, bills of materials, assembly instructions
Production	Customer feedback, retail data	Sales presentations, demonstrations, photographs, product instructions, presentation graphics

Data from Ulrich & Eppinger, 2011.

TABLE 3.3 *Information Use in the Engineering Design Model of Dym and Little*

Design Stage	Sources of Information	Outputs
Problem definition	Client's statement; literature on state-of-art, experts, codes, and regulations	Revised problem statement; detailed objectives, constraints, user requirements, and functions
Conceptual design	Competitive products	Conceptual design solutions; design specifications
Preliminary design	Heuristics, simple models, known physical relationships	Selected design solution; test and evaluation methods
Detailed design	Design codes, handbooks, local laws and regulations, suppliers' component specs	Proposed fabrication specs; final design solution for review
Design communication	Feedback from customers, required deliverables	Final report to client containing fabrication specs and justification for those specs

Data from Dym & Little, 1999.

understanding the problem, understanding constraints, and communicating (all of which have information-based components). Gathering information came out ahead of analyzing, brainstorming, planning, prototyping, testing, and building, for example. Atman et al. (2007) found that, with experience, engineers make increasing numbers of information requests when solving design problems. The number of sources, kinds of requests, and time spent gathering information all increased substantially when comparing groups of first-year, senior, and professional engineers. Bursic and Atman (1997) also found a positive correlation within each group between the number and kinds of requests and the quality of the final products, although they believed that even the senior students needed substantial improvement in their use of information in the design process.

Several studies of user behaviors have attempted to quantify the impact of information on success for engineers. Tenopir and King (2004) studied the habits of university and national laboratory engineers and scientists and found that university engineers read on average twice as many articles as engineers at national laboratories. In terms of time, engineers spent about 90 hours a year, or 5 percent of their time, reading journal articles. Overall, engineers reported spending 280 hours per year reading some form of documents, more than they spent in informal discussions (104 hours) or internal meetings (136 hours). They also found that engineers who had won awards or received other recognitions of excellence read on average about twice as many articles as those who didn't. Many corporations have gatekeepers—that is, engineers who are more familiar with information resources, including a network of professional contacts, and who are often the go-to people for help answering information needs. These gatekeepers tend to publish more than their counterparts, and their employees tend to perform better than the company average.

Engel, Robbins, and Kulp's (2011) survey of engineering faculty at 20 different institutions found that more than three quarters reported seeking information at least weekly to prepare for student lectures, and over half reported seeking information at least weekly both for their research projects and to stay current in their field. According to this survey, engineering faculty about equally often use conferences, current journals, personal communication, and following article references as ways to stay abreast of developments in their field. They still rely on discussions with colleagues and students as significant sources of information, but they rely even more on scholarly journals and Internet resources, with monographs and conference attendance rated highly, although not quite as highly as discussions. Engel, Robbins, and Kulp (2011) found ease of access the most important factor for engineering faculty when gathering information; therefore, electronic access to current and historical journals were of primary interest, although print books were still rated more highly than e-books in importance by respondents. Kwasitsu (2003) found that practicing engineers with an advanced degree used scholarly literature more frequently than did those without, implying that the increased familiarity with those sources might make them more accessible to those engineers in the workplace.

INFORMATION HABITS OF ENGINEERS

Studies have consistently found that engineers engage in information activities for on average between 20 and 40 percent of their workday, which is more time than they spend on more traditional design activities such as prototyping and modeling (Tenopir & King, 2004). Information activities here include locating, using,

producing, and communicating information in any format. Characterizing the information habits of engineers can be problematic, however, since they may take on a wide variety of roles within a project team, and there are substantial disciplinary differences between information use habits. As Tenopir and King (2004) indicate, during his or her career, an engineer may assume a variety of functions, "including research and development, design, testing, manufacturing and construction, sales, consulting, government and management, and teaching" (p. 78). They go on to state that, for example,

> design engineers want original, up-to-date information, relying heavily on internal reports and test results rather than the published literature. In a consulting role they rely more on external market information about vendors and customers. When an engineer takes on an administrative role, he or she needs a wider variety of both external and internal information, including regulations, information on new technologies, personnel records, and business information. R&D information needs similarly vary with each stage of the project. (p. 79)

That said, some general principles can be drawn. As Leckie, Pettigrew, and Sylvain (1996) found, engineers, like other professionals such as health care workers and lawyers, engaged in very context-specific information-seeking behaviors and rely heavily on their previous knowledge and personal collections when approaching a problem. Overall, engineers' information-seeking behaviors have consistently been characterized as a least effort approach. That is, engineers act in a way to minimize the work involved when searching for information, rather than to maximize the results of the search. Engineers will accept a lower quality information source if it is easier to locate, access, and/or apply to a problem, with Gerstberger and Allan (1968) finding no correlation between source quality and use. Kwasitsu (2003) found that quality, relevance, currency, and reliability of the information source ranked significantly lower than accessibility and availability, although they all were rated as important by the majority of the engineers surveyed.

Thus traditionally, colleagues and personal collections have provided the lower barrier to locating information, and engineers will use their personal collections preferentially even though they might be of limited scope. However, gathering information from colleagues is not without drawbacks, as the time spent locating an appropriate colleague, the intellectual and social effort involved in interacting, lack of specificity of answers, poor memory of their subjects, and inappropriate information have been described as challenges (Tenopir & King, 2004). Furthermore, some engineers are intimidated by admitting to a colleague their ignorance on a subject. Although colleagues and personal collections traditionally have been preferred, recently, Googling has become a first-resort method of locating information for engineers as well (Allard, Levine, & Tenopir, 2009; Hirsh & Dinkelacker, 2004).

Hertzum and Pejtersen (2000) investigated the social aspects of information seeking and found that the search for documents and people is frequently intertwined. Since technical documents are static, when more context is desired, engineers go to the human source of the information, especially to explain how results can be appropriately applied to a problem or to interpret the information implicit or missing from the document. By consulting a trusted expert, engineers also frequently gather feedback on their own ideas. Conversely, technical documents contain specific facts and figures, and since memories fade with time, having access to those pieces of data provides a level of assurance of the accuracy of the information. Often, the process is iterative, with engineers finding people who know where the useful documents are and what they contain, and documents in turn providing pointers to experts who can expand on a particular topic. As a rule of thumb, the more complex, uncertain, or ambiguous the task, the more likely an engineer will search out a personal contact instead of a documentary resource. With the growth of the Web, including videos, tutorials, and forums, richer information can be made available without contacting colleagues directly, so the balance of personal and documentary information gathering is changing as well.

Ellis and Haugan (1997) explained different information habits based on the type of problem faced. They classified problems as incremental, radical, or fundamental. Incremental projects primarily involved conversations with colleagues to understand the context for minor improvements to a product. Radical projects involved major redesign of a product or service. In these cases, collegial interactions are supplemented with environmental scanning of current technologies or principles, mainly through reading review articles and conference proceedings. Fundamental projects are those in which a company moves into a completely new area. Since there will be little in-house expertise in this kind of project, engineers typically begin with a literature review before consulting others. This kind of activity requires the most in-depth information seeking and is most likely to include consultation with corporate librarians and use of formal library materials.

In terms of the actual kinds of textual resources accessed by engineers, corporate intranets that contain internal reports and data dominate the usage. Journals and conference proceedings, patents, marketing data, regulations, standards, external technical reports, and product information also are common information sources. Depending on the role of a particular engineer or the field he or she is working in, the distribution of sources varies significantly. Research and development engineers, for example, have a profile of information use similar to scientists, while production engineers or marketing specialists will have utilitarian information needs.

Jeffryes and Lafferty (2012) surveyed returning co-op students, largely mechanical engineers, as a proxy for entry-level engineers and found that, in their internships, 75 percent used standards, 60 percent used books, over 50 percent used technical reports, 33 percent used journal articles, and 20 percent used patents, and the vast majority learned how to locate all those information sources except books during their college career.

Generally speaking, engineers dislike searching for information in the typical indexes that librarians love. Rather, most engineers locate information through recommendations from colleagues or citations from other papers, or as a result of their own current awareness browsing of technical or trade journals, blogs, and so forth. Tenopir and King (2004) found that about half of journal articles read by engineers in their study were located through browsing, with another third coming as suggestions from colleagues. Only 10 percent of papers read were located through conscious searching. Again, as Internet search engines have substantially decreased the barrier to searching, information habits are changing.

BARRIERS TO INFORMATION USE

As mentioned in the previous section, engineers tend to take a least effort approach to information gathering. Several factors can contribute to increasing the effort of searching, including the fiscal and psychological cost, accessibility of

resources, lack of familiarity with appropriate sources, inappropriate formats, irrelevance, and lack of high-quality material.

Cost

Costs come in different forms, with monetary costs actually influencing engineers' behaviors least. Rather, time is the most important cost, including the time it takes to search, acquire, and process the material. Additionally, the mental cost—that is, devoting one's attention to the process of finding information—is another important component.

Accessibility

Does an information source exist and is it available to be accessed? Again, there can be many levels of accessibility. In the past, a physical journal might have been located in a locked library after hours. Now, the information might exist, but it could be behind a subscription wall (and although the monetary cost might not be a barrier, the process of acquiring access could be). An information source might exist but be buried in a poorly constructed knowledge management system, so therefore inaccessible to the end user. Gerstberger and Allan (1968) found that the more experience an engineer had with a particular information source, the more accessible he or she found it to be.

Familiarity

Lack of familiarity with a resource type or information system also leads to nonuse. In line with the principle of least effort, if a search system is unfamiliar, it will take much more effort to use effectively. Similarly, if an engineer has not used patents, standards, or technical documents before, or has not heard of a particular collection of documents, these are not in that engineer's toolbox of sources and thus will be neglected in the search for appropriate information.

Format

An information source might contain appropriate content to solve a problem, but it might not be in a format usable by the engineer. For example, the treatment of the topic might be at a level inappropriate for the background of the reader. Alternatively, the method of encoding the information (textual, graphic, or electronic) might not allow for easy importing into a project. Data files might be in a different format than that used by a project's software programs, or perhaps the project team needs a drawing, when only a written description is available. Engineers determine whether it is worth their time and effort to convert information into a usable format.

Relevance/Information Overload

When conducting searches, engineers struggle with sifting through an overwhelming number of results, most of which are not relevant to their search. Engineers often consult with colleagues to locate relevant information, whether internal or externally produced documents, as well as for assistance with extracting the appropriate information from those documents and with the context of the application of that information.

Quality

Engineers desire high-quality information, and although quality doesn't rank as the most important factor, it does rank highly in their search process. The difficulty is locating high-quality information and determining which information is of high quality. Particularly since engineers tend to have little patience with searching specialized databases, including, frequently, corporate intranets, they may only be looking at the open Web, excluding many high-quality sources from their searches. Furthermore, engineers at smaller firms often do not have ready access to subscription material such as journals, further limiting their ready access to high-quality materials.

SUMMARY

The previous discussion indicates that the information-seeking behavior of engineers is quite complex but that, overall, the more advanced and accomplished an engineer, the more information the engineer seeks and uses in his or her professional career. While engineers prefer finding information from their personal collection and from their colleagues, they increasingly rely on Internet search engines. When they need accurate facts and figures, they do consult the written record, whether internally or externally produced. Information habits center around the concept of minimizing effort, rather than maximizing the value of information retrieved.

Increasing the effectiveness of engineers' information-seeking habits, then, requires a combination of training, to increase the familiarity and accessibility of resources, and improvement of knowledge management systems, to increase accessibility of previously located resources. Learning about different document types (e.g., technical reports, patents, journal articles), as well as search systems, will enable engineering students to conduct, in terms of time and effort, a lower cost search for information. Students need to be trained to extract information efficiently from different resources—for example, to read a scientific paper effectively and to become familiar with sources that provide information in a variety of formats (e.g., tabular, graphical, textual)—so that the information is not only available but usable in the context of the problem at hand. Finally, in order for engineers to develop their own personal knowledge bases, training in knowledge management tools and the habits of using them are critical so that information doesn't become forgotten or lost to the system. Since engineers almost exclusively resort first to their personal collection of information, the better their knowledge management system, the more effective they will be in their careers.

All of these information literacy principles—locating, accessing, using, and learning from information—need to be instilled in engineering students so that they can thrive in their increasingly competitive knowledge-based society. In order to achieve this goal, we have developed an information-integrated model of engineering design, which is introduced in the following chapter.

REFERENCES

Allard, S., Levine, K. J., & Tenopir, C. (2009). Design engineers and technical professionals at work: Observing information usage in the workplace. *Journal of the American Society for Information Science and Technology*, 60(3), 443–454. http://dx.doi.org/10.1002/asi.21004

Atman, C. J., Adams, R., Cardella, M., Turns, J., Mosberg, S., & Saleem J. (2007). Engineering design processes: A comparison of students and expert practitioners. *Journal of Engineering Education*, 96(4), 359–379. http://dx.doi.org/10.1002/j.2168-9830.2007.tb00945.x

Bursic, K. M., & Atman, C. J. (1997). Information gathering: A critical step for quality in the design process. *Quality Management Journal*, 4(4), 60–75.

Dym, C. L., & Little, P. (1999). *Engineering design: A project-based introduction*. New York: Wiley.

Ellis, D., & Haugan, M. (1997). Modelling the information seeking patterns of engineers and research scientists in an industrial environment. *Journal of Documentation*, 53(4), 384–403. http://dx.doi.org/10.1108/EUM0000000007204

Engel, D., Robbins, S., & Kulp, C. (2011). How unique are our users? Comparing responses regarding the information-seeking habits of engineering faculty. *College & Research Libraries*, 72(6), 515–532.

Gerstberger, P. G., & Allen, T. J. (1968). Criteria used by research and development engineers in the selection of an information source. *Journal of Applied Psychology*, 52(4), 272–279. http://dx.doi.org/10.1037/h0026041

Hertzum, M., & Pejtersen, A. M. (2000). The information-seeking practices of engineers: Searching for documents as well as for people. *Information Processing and Management*, 36(5), 761–778. http://dx.doi.org/10.1016/S0306-4573(00)00011-X

Hirsh, S., & Dinkelacker, J. (2004). Seeking information in order to produce information: An empirical study at Hewlett Packard Labs. *Journal of the American Society for Information Science and Technology*, 55(9), 807–817. http://dx.doi.org/10.1002/asi.20024

Jeffryes, J., & Lafferty, M. (2012). Gauging workplace readiness: Assessing the information needs of engineering co-op students. *Issues in Science and Technology Librarianship*, 69. http://dx.doi.org/10.5062/F4X34VDR

Kwasitsu, L. (2003). Information-seeking behavior of design, process, and manufacturing engineers. *Library and Information Science Research*, 25(4), 459–476. http://dx.doi.org/10.1016/S0740-8188(03)00054-9

Leckie, G. J., Pettigrew, K. E., & Sylvain, C. (1996). Modeling the information seeking of professionals: A general model derived from research on engineers, health care professionals, and lawyers. *The Library Quarterly: Information, Com-*

munity, Policy, 66(2), 161–193. http://dx.doi.org/10.1086/602864

Mosberg, S., Adams, R., Kim, R., Atman, C. J., Turns, J., & Cardella, M. 2005. Conceptions of the engineering design process: An expert study of advanced practicing professionals. In *Proceedings of the 2005 American Society for Engineering Education Annual Conference & Exposition.* Washington, DC: American Society for Engineering Education.

Tenopir, C., & King, D W. (2004). *Communication Patterns of Engineers.* Hoboken, NJ: Wiley.

Ulrich, K. T., & Eppinger, S. D. (2011). *Product Design and Development* (5th ed.). New York: McGraw-Hill.

Wodehouse, A. J., & Ion, W. J. (2010). Information use in conceptual design: Existing taxonomies and new approaches. *International Journal of Design, 4*(3), 53–65.

CHAPTER **4**

INFORMATION-RICH ENGINEERING DESIGN

An Integrated Model

David Radcliffe, Purdue University

Learning Objectives

So that you can implement a robust, informed approach to teaching design to students, upon reading this chapter you should be able to

- List and describe the seven essential activities of engineering design used to frame this book
- List and describe the major information-seeking/ creating activity associated with each of the seven elemental design activities
- Characterize the seven major information-seeking/ creating activities associated with each of the elemental design activities in terms of variety and depth of information required
- Outline the implications for mapping potentially helpful information literacy interventions in engineering design courses

INTRODUCTION

The review of the nature of engineering design in Chapter 1 revealed a many-faceted, contingent, sociotechnical endeavor that is difficult to define, capture, and characterize in a simple manner. While recognizing the complex, emergent nature of engineering design and the diversity of perspectives, for the purposes of this handbook we have distilled from the analysis in Chapter 1 seven elemental activities that are part of any engineering design project. These are not intended to be a linear prescribed set of actions in an engineering design project. On the contrary, most of these activities occur at multiple times across any project, perhaps at different scales and at different levels of detail. For instance, one of the seven activities involves organizing a project team. While this happens initially at the beginning of a project, inevitably there are changes in the composition of the team in terms of personnel and roles related to changes in emphasis and disciplinary expertise as a design project unfolds. In that way aspects of team formation can occur at multiple points throughout a project.

These seven elemental activities are not another model of the engineering design process. They are offered simply as a convenient device for organizing the material in this handbook—a placeholder for whichever conception or model of design a particular educator or academic tradition prefers to use when introducing students to engineering design—and are used to focus attention on the different types of information-related activities that engineering students should master. The intention is that the ideas around information literacy pertinent to each of the seven design activities can be readily mapped back by the reader to any particular model of engineering design.

ELEMENTAL ENGINEERING DESIGN ACTIVITIES

Of the seven elemental engineering design activities considered in this framework, five reflect the recurring ideas from the descriptive and prescriptive models: *clarify* the task, *synthesize* many possible solutions, *select* the most suitable solution, *refine* the preferred solution, and *communicate* the solution to inform and persuade the stakeholders. The other two activities acknowledge the social dimension of design: *organize* the team, and throughout the design project continuously *reflect* upon and improve processes. These activities are represented in Table 4.1.

These activities cover the product development process up to the point where the proposed solution is documented such that it can be made and implemented. Of course, the complete life cycle of a new product, system, or service includes the subsequent processes of manufacture, installation, commissioning, operations, maintenance, updating as technologies change, retirement from operation, and reuse or recycling of the component elements (McDonough & Braungart, 2002). The whole life cycle also includes such things as the training of users or operators or other service or support staff and provision of necessary support infrastructure and spare parts.

Decisions made in these early stages of the product realization process shape the subsequent or downstream life stages, including such things as the whole of life cost of the product, system, or service being designed and its overall sustainability. Thus, the earlier relevant information is introduced, the larger its impact on the entire product life cycle;

TABLE 4.1   *Summary of Elemental Engineering Design Activities*

	Design Activity	Example Tasks
Improve Design Work Processes — Reflect on and analyze what is working and what needs improvements throughout the project — Use a disciplined mode of reflection to capture lessons learned	*Organize Your Team*	Select/change team members to achieve a diversity of knowledge, skills, and qualities Agree on a code of conduct and active modes of intra-team communication Build/renew team cohesion including a shared understanding of team dynamics Adopt team maintenance tools and process improvement schedule Establish information strategy, including capture, storage, and use Define team member roles, reporting, and review processes Review and improve processes throughout the project
	Clarify the Task	Analyze the brief and any other initiating documents Speak with client, potential users, and other key stakeholders; ask questions Identify additional sources of information that will establish the wider context Estimate the order of magnitude of things associated with the project Develop a list of possible risks and opportunities Determine the scope of the work to be done in relation to the wider context List the requirements and constraints for the product/system/process Articulate the specific design criteria/measures of success
	Synthesize Possibilities	Explore the *prior art* in the widest sense of the term Investigate similar and quite different operational contexts for ideas Gather information for existing artifacts, literature, experts, observation Develop as many concepts and combinations of concepts as possible Test ideas and improve initial concepts to learn more about the tasks Refine scope of work; relax constraints
	Select Solution	Select the most promising concepts from the many options Flesh out (embody) preferred concept(s) and analyze these to understand performance Conduct a design review with client based on this analysis and gain feedback
	Refine Solution	Visualize/model the manufacture, installation, operation, and maintenance Estimate cost structure for whole of life cost Refine risk and opportunity analysis
	Communicate Effectively with all Stakeholders	Identify and stay in regular communication with key stakeholders who need to be heard, informed, or persuaded at any time during design process Get to know and appreciate their perspective and hence their information needs

hence the critical importance of integrating information literacy (broadly defined) as early as possible into the design process and blending it into the education of engineering students as they learn to think as engineering designers.

INFORMATION-RICH ENGINEERING DESIGN (I-RED) MODEL

During engineering design, existing information is used and new information is generated. In this handbook the shorthand term *information-seeking/creating* is used to capture this idea. Figure 4.1 outlines the seven information-seeking/creating activities associated with each elemental design activity.

Organize Team: Develop Knowledge Management Strategy

In forming and/or modifying a design team for a particular project or phase of a project, the goal is to gather the most appropriate range of disciplinary backgrounds with sufficient levels of knowledge and experience and complementary personal attributes and professional skills. Factors that influence team performance include the range of technical knowledge and skills, temperaments, work styles (e.g., starters versus finishers, big-picture versus detail-oriented people), organizational and leadership skills, and oral and written communication skills.

From an information-seeking/creating perspective, the primary objective in *organizing the team* is to develop a strategy for organiz-

Engineering Design Activity	Information-Seeking/Creating Activity	
Organize Your Team	*Develop Knowledge Management Strategy* How will the team acquire and manage information?	
Clarify the Task	*Establish the Project Context* What do the stakeholders want and what are the constraints?	
Synthesize Possibilities	*Investigate Prior Art* What have others done in similar situations?	*Improve Knowledge Management Processes* *How do we capture and use lessons learned?*
Select Solution	*Assess Technologies and Methods* How do the solutions actually work? How can components work together?	
Refine Solution	*Integrate Technical Details* What detailed technical information is available?	
Communicate Effectively with all Stakeholders	*Distill Project Knowledge* What is the critical information that must be passed on?	

Improve Design Work Processes

FIGURE 4.1 Information-seeking activities corresponding to design activities (Table 4.1 and Figure 4.2).

ing and managing information. It is imperative that the strategy and the metadata structures and tools to be used for knowledge management throughout the project be carefully thought through before the project work commences. This is an upfront investment that can pay significant dividends later in the project in terms of effort saved by not wasting time locating project-critical information, ensuring that ideas and information from an early phase of the project are not forgotten by a later stage, and expediting and making the intermediate and final communications and documentation of project information much more efficient and effective.

One set of skills often overlooked when considering a knowledge management strategy is the level of information literacy of the members. By including team formation as part of the I-RED model, attention is focused on the need to establish a core capability amongst the members to be able to identify, locate, gather, analyze, synthesize, and share information (information seeking) within the team and with other stakeholders. The information literacy of the team sets a foundational baseline in terms of the ability of team members to seek and share information effectively, which in turn is a key determinant of the overall effectiveness of the design work they undertake.

Clarify Task: Provide Context

The team's purpose in *clarifying* the task is to create a coherent and cogent description of purpose and scope of the design need or opportunity before them. This includes establishing a set of criteria by which the outcome will be judged by the client or user and the other stakeholders more generally. The client might give an initial need statement, such as, "I need a water purification system for a community of 2,000 people." From that brief statement, the team must determine what specific objectives the client may have, quantify and clarify the specific requirements, and determine the constraints or opportunities, including the type and amount of resources available for the solution. Much of this phase involves working with clients to better understand their expectations. Sapp Nelson (2009) found that the library science technique of reference interviewing can facilitate better elicitation of client requirements. Clients and/or users often state their need in terms of a solution. The design team has to unpack this to identify the underlying need that must be satisfied.

This design activity centers on gathering preliminary information on the broad context of the design task. In the case of the water purification system example, this might include exploring the different types of purification systems, specific health risks of unclean water, and the local cultural, economic, and political environment of the stakeholders. Seeking out such information can help the team generate pertinent questions for the client and other stakeholders to help them articulate objectives they didn't realize they had and to surface constraints or conditions that will limit or bound the possible solution space. These questions are an instance of information creation. If there are regulations or other legal requirements—for example, clean water standards—then those are constraints on any solution.

Information seeking during these activities centers on general sources of information, such as encyclopedias, trade magazines, or handbooks, which can give an overview of the major technologies being used to solve the problem. Codes and regulations provide guidance on legal constraints. When teaching the informational component of this phase, focusing on the initiation stage of the Information Search Process (ISP) is the most important (Kuhlthau, 2004). This is the phase when students will need to determine what information they know and what information they still need to find. Often with novices, "they don't know what they don't know," so they have difficulty articulating the need for information. Providing students with some structure for asking questions can facilitate their moving beyond an "ignorance is bliss" phase and get them to engage with what they don't know.

Synthesize Solutions: Investigate Prior Work

During this design activity the team consolidates and prioritizes a list of design requirements uncovered during task clarification and they explore potential design solutions that could meet those assembled needs and any constraints. This is a very creative phase, involving brainstorming and other activities focused on idea generation and the *synthesis* of possible solutions. As such there is a considerable amount of new information generated that has to be organized and managed lest good ideas get lost. A valuable trigger for this is to explore the prior art, solutions to similar problems that others have designed, and other technologies that might have novel applications to this problem. In order to enlarge the range of potential options to the fullest extent possible, an eclectic range of information types and sources need to be consulted. While the patent literature might be the most obvious source of information on specific technologies, at this phase of the process, where the emphasis is on developing a large number of possibilities, a more efficient way to investigate prior art might be to peruse the popular literature for reports of other solutions, including material provided by engineering firms, nonprofits, or other organizations that have worked on similar problems.

As part of creating options, the design team needs to consider the whole life cycle of potential solutions. This can include considerations of how to build it, how it will be used after fabrication, how it will be maintained, and what will happen when it reaches the end of its life cycle (repurposing, reuse, or recycling, for example).

Select Solution: Assess Technologies and Approaches

This design activity is where conceptual design solutions are evaluated to determine which solution will finally be *selected* for detailed development and implementation. This can involve selecting a short list of two or three prospective concepts from a larger initial set of ideas and approaches. The final selection of the most suitable concept usually requires that the two or three prospective concepts be fleshed out (embodied) in the form of basic configurations that can be evaluated—for instance, as a computer model to determine whether these preliminary design concepts are feasible and practical. Often this is a hands-on phase of design, during which the team makes simple or more sophisticated prototypes and conduct tests to see if they meet the design specifications. So as to facilitate testing of the ideas, an overall system might be decomposed into a series of subsystems that can be evaluated. In that case, the inputs and outputs of each subsystem will have to be determined to ensure compatibility and interoperability. Again there is a considerable amount of information generated during this design activity.

For this phase, standard testing processes, laboratory and experimental procedures, and information about appropriate simulation/modeling software could all be needed. This enables the team to determine whether a particular model is appropriate for the use case of the design problem, and whether, for example, the results can be extrapolated from a model to the full scale. Additionally, the management of new data and information assembled and created during prototyping and testing needs to be carried out appropriately. As Carlson, Fosmire, Miller, and Sapp Nelson (2011) note, data information literacy is a robust new area for librarians to apply (and teach) information management skills to the curation of data.

Refine Solution: Assemble Detailed Technical Information

The focus in *refining* the solution is on developing and documenting an increasingly detailed description of precisely what the product, system, or service will be like. This is an information-intensive activity wherein the selected preliminary design is turned into something that can actually be made. For example, the actual materials or standard components to be used are selected to ensure that they all meet the relevant codes and regulations for performance. Questions such as will pieces fit together, can the component be serviced without taking apart the entire artifact, and can the output of one stage of the artifact be used as an input in the next stage are all important to resolve. Such considerations apply not only to hardware but also to software. For example, writing computer code for a software program involves the construction of modules and objects, many of which may come from preexisting standard libraries. As a result, it is very important that the output of an object is in a format and with appropriate units that can be used in a subsequent routine.

For this design activity, handbooks, product catalogs, and component specifications are all important to make sure that the result is practical and achievable. Patents will shed light on the more cutting edge technologies that could be licensed for use in the project.

Communicate: Distill and Translate Project Knowledge

The completed description of the product, system, or service needs to be communicated to those who will make it, install it, operate it, maintain it, update it, and even dismantle and recycle components of it. The amount of information necessary to describe all these facets of even a relatively simple product is substantial. For a large system the quantum is enormous. The nature and the format of the information that is required for all the stakeholders is significantly different than the core technical information necessary to define the product, system, or service that was designed. New information based on this core technical data must be generated in order to interpret the core description to particular audiences. For instance, much of the information in a user manual is not developed explicitly as part of the creation of the core technical description. The user manual draws on this core description and many explicit and some implicit assumptions that went into a variety of design decisions made throughout the project. The relevant information has to be *distilled* and then *translated* into a form and a format that makes it easily accessible to the user. The same applies for the additional information needed to guide the manufacture, assembly, installation, operation, and maintenance of the product, system, or service.

This process actually takes place as part of each of the forgoing design activities and not simply at the completion of detailed design. By communicating ideas and partial details and seeking feedback from the relevant stakeholders throughout the entire project, the design team can much more effectively manage expectations and identify potential problems early and remedy them before too much time or resources have been expended on an idea or a detail that will ultimately not succeed.

Thus the design team should capture the information found and generated during each design activity, including any computer models and modeling data, tests plans and data, mock-ups, functional prototypes, and the like. It is especially important at this point that information is well documented. Others will be using the information presented in this section, so they need to know where information exists, for example, on the safety codes for operation, or the material composition of components for potential recycling. The most recent and complete information about supplier information, codes met, availability of replacement parts, or authorized maintenance all are important in the final documentation.

Reflect: Organize and Document Lessons Learned

Throughout all the design activities the team must strive to improve their processes of working to be more holistic, more effective, and more efficient. Central to this is continuously improving their knowledge management processes and being disciplined and diligent in staying up to date with their information-seeking activities.

During the clarification of the task many types of information are gathered, and it is often difficult to know with any certainty

which ones are going to be particularly useful later in the project. Time spent organizing and curating early information, much of which may turn out not to be important, can prove to be wasted once the direction and scope of the project becomes clearer. Equally, not capturing and describing this early information could prove very costly later. There is no simple solution to this dilemma; each project has a unique set of problems of this type. One effective approach is to *regularly use* the knowledge management strategy developed as part of organizing the team and to learn from that experience. The system should be periodically reviewed and improved as the problematic issues around information handling in the particular project reveal themselves.

Taken together, the series of elemental design activities and corresponding information-seeking activities comprise the I-RED model, depicted in Figure 4.2

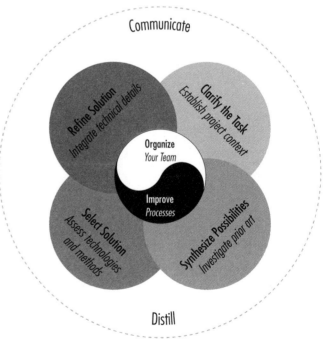

FIGURE 4.2   Information-Rich Engineering Design (I-RED) model.

PROMPTING QUESTIONS FOR INFORMATION-SEEKING ACTIVITIES

Each of the information-seeking/creating activities is characterized by a series of prompting questions, as shown in Table 4.2. This aligns with the notion of design as a question-asking process (Eris & Leifer, 2003). Pilerot and Hiort af Ornäs (2006) follow a similar approach in formulating guiding questions from not only a process- but also a product-oriented perspective. At a macrolevel the overall trend in information seeking/creating follows the ISP stages. Within each information-seeking activity corresponding to an engineering design activity, the ISP moves from exploration within uncertainty toward a focus on more pertinent information that defines the later part the activity. As a project proceeds, the members of the design team tend to follow those stages described by Kuhlthau (2004)—that is, they go from uncertainty, to optimism, to confusion and doubt, which gives way to greater clarity and a sense of direction leading to, hopefully, satisfaction and accomplishment.

MAPPING I-RED ACTIVITIES TO INFORMATION SPACE

The six pairs of engineering activities and information-seeking/creating activities at the core of the I-RED model can be located in an information space with orthogonal axes for the variety of knowledge domains and the level of specialization in a given domain. This is illustrated in Figure 4.3.

The location of each activity bubble indicates the relative breadth and depth of the types of information sought/created in the corresponding design activity. The engineering design activity of reflection on processes and the corresponding information-seeking/creating support activity of managing information and documenting learnings occur throughout all other activities. This is depicted in Figure 4.3 as a substrate (the blue ellipse) to indicate that these are pervasive activities that underpin all the others and also links them. The arrows between activities indicate that information is passed on from one activity to another.

By its location in the information space, the *organize team/develop knowledge management strategy* activities draw on a reasonable diversity of knowledge domains and an intermediate depth of specialization. However, the activities around *clarifying the task/providing context* by necessity draw upon a very diverse range of knowledge domains, although the depth of knowledge in each is relatively shallow, at least initially. Knowledge of the relevant context increases as the concepts are developed, selected, and detailed. Seeking information around prior work to support the synthesis of many possible solution concepts is more focused in terms of the variety of knowledge domains but correspondingly deeper in terms of the level of specialization. This is so because the task clarification process has reduced the scope of possibilities.

This trend of there being fewer knowledge domains yet more depth of knowledge and specialization of information type continues through the selection of suitable solutions by assessing various approaches and technologies and refining the preferred solution through gathering together substantial amounts of relatively specialized technical information.

However, in order to communicate the large amount of information that defines the final product, system, or service that was designed back to a variety of stakeholders, including the client and/or user, this information has to

TABLE 4.2 *Example Prompting Questions for Each Information-Seeking Activity*

Information-Seeking Activity	Example Prompting Questions
Develop knowledge management strategy	What is the level of specialization and variety of technical and other knowledge across the team members?
	What is their level of proficiency in information seeking and critical evaluation?
	What additional information-seeking skills are required? How might additional information skills be best developed?
	How will they develop and implement communication and documentation policies and infrastructure?
Establish the context	What are the historical, social, cultural, political, geographical, and economic contexts of the problem?
	Who are the stakeholders? Who will use this product, system, or service throughout its life cycle—from the cradle to the grave?
	What are the most important requirements or functions for various stakeholders?
	What is absolutely necessary (needs) and what is discretionary (wants)?
	What are the measures of success from the perspective of all stakeholder groups?
	What codes or regulations does the end system/product have to comply with?
Investigate prior work	What approaches are possible to address this type of problem?
	What examples of solutions exist for this type of problem?
	What existing products, systems, or processes tackle similar needs or opportunities?
	What technologies might be used to tackle this need or opportunity?
Assess technologies and methods	How do the technologies scale from a prototype to full-scale implementation?
	How would different specifications of performance be tested?
	Are there relevant standards for conducting tests of materials or components?
	What tools would help in designing a full-scale model? What modeling or design software do professionals use in this field?
	What benchmarking information is available for competing products?
	How do proposed new solutions compare to existing ones in terms of performance, user desirability, financial viability, or other indicators of success?
	How can the quality of externally provided information be assessed?
	How do the technologies work at a deep level? What are the inherent strengths and limitations of the technologies?
	What is required to create, operate, and maintain these technologies?
Integrate technical details	What properties does a component have and what does it need to have to work properly within the system?
	What components need to be fabricated, and what properties do they need to have to work with the rest of the system?
	What components already exist that can be used as part of the solution?
	What are the standard inputs/outputs for the systems or subsystems (e.g., appropriate interfaces, size of conduits for moving materials)?

TABLE 4.2   *Example Prompting Questions for Each Information-Seeking Activity*—cont'd

Information-Seeking Activity	Example Prompting Questions
Distill and translate project knowledge	Is the documentation prepared and presented in a form and style most appropriate to the future user of that information? What are the most important ideas and details to present to particular stakeholder groups? Why? How can this best be done?
Improve knowledge management processes	What new information was generated and how important or valuable is it? Has all the pertinent information gathered/created and used in the design process been fully documented and cataloged, including calculations, models, graphic images, tables, and other non-textual information? Are all stages of the product/system/ process life cycle adequately documented?

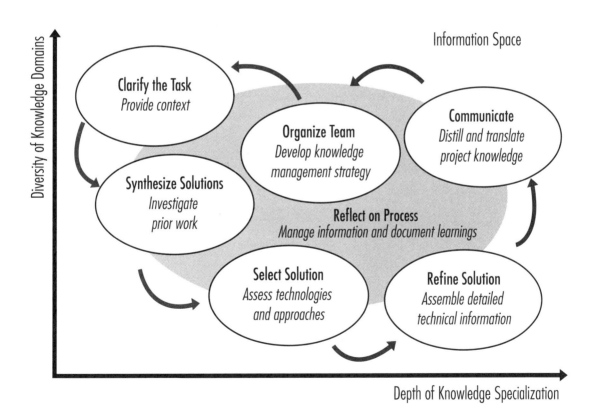

FIGURE 4.3   Mapping Information-Rich Engineering Design (I-RED) activities.

be distilled and translated into forms that are suitable for a wide range of people who think, work, and live in a diverse range of knowledge domains. Thus, this set of activities is shown at the top right of the information space, indicating that it involves in-depth and specialized information that must be understood in quite different knowledge domains.

APPLICATION OF THE I-RED MODEL

The I-RED model provides a descriptive rather than prescriptive approach to identifying how and when information-seeking/creating activities and training in information literacy can be integrated into the engineering design process. Both the informational and engineering design components are described as generally and generically as possible so that the model can be applied to a wide range of engineering disciplines. The purpose is to step outside of the jargon of both library science and engineering design to enable practitioners on both sides to talk directly and productively about student and project needs. The motivating factor of the model is for students to be able to determine at each stage what information they need at that time to move the project forward and how they can acquire and use that information. Instead of requiring students to do a literature review at the beginning or end of a design project, this model provides guidance for information gathering activities that can continue throughout the life of the project. This should provide students with the ability to take an integrated approach that will enhance the richness of the design of the final artifact.

This model captures the idea that as a *learning process* design creates knowledge as well as consumes it. Thus the members of the design team contribute to the body of knowledge.

In industry this new knowledge would likely appear in a corporate intranet or knowledge management system. Historically, such new knowledge has been poorly managed in student design project teams, in part due to the lack of easy to learn and use knowledge management systems that scale to projects that may last one semester and involve a team of only five or six students. However, with the advent of large scale, lengthy student-led projects— for example, vehicle projects or service projects that extend over multiple years, during which the membership of a team might change every semester or year—much more effective knowledge management systems are needed.

The type and scope of information sought and generated in engineering design activities is very broad. Design information is not limited to documents such as books and catalogs, whether in physical or electronic form. It also comprises still and moving images; multidimensional datasets, including product and geographical information; the spoken word; and physical and virtual artifacts. The sources for and modes of gathering, capturing, analyzing/interpreting, storing, and sharing this eclectic range of information is enormous and ever changing. This has critical implications for both the development of information literacy skills in students and the work of university librarians who support design projects in engineering schools.

SUMMARY

The I-RED model combines conceptions of the design process and information literacy to create a logical framework for integrating the development and use of information skills into engineering design course work. This model also draws on my experience of teaching engineering design over many years in both the United States and Australia, including numerous collaborations with librarians to embed instruction on information literacy within the design curriculum.

With this conceptual model under our belt, the next question is how to implement these principles. The rest of this handbook investigates the main information activities corresponding to the general steps of an engineering design process model. The I-RED model is not expected to replace whatever engineering design model you may be currently teaching your students. Rather, I-RED can be integrated into your preferred models. The following chapters provide examples of activities that can be easily incorporated in a design course, with the rationale for why these information steps are important and necessary, and the resources to carry out the instruction.

REFERENCES

Carlson, J., Fosmire, M., Miller, C. C., & Sapp Nelson, M. (2011). Determining data information literacy needs: A study of students and research faculty. *Portal: Libraries and the Academy, 11*(2): 629–658. Retrieved from http://muse.jhu.edu/journals/portal_libraries_and_the_academy/v011/11.2.carlson.pdf

Eris, O., & Leifer, L. (2003). Facilitating product development knowledge acquisition: Interaction between the expert and the team. *International Journal of Engineering Education, 19*(1), 142–152.

Kuhlthau, C. C. (2004). *Seeking meaning: A process approach to library and information services* (2nd ed.). Westport, CT: Libraries Unlimited.

McDonough, W., & Braungart, M. (2002). *Cradle to cradle: Remaking the way we make things.* New York: North Point Press.

Pilerot, O., & Hiort af Ornäs, V. (2006, August). *Design for information literacy: Towards embedded information literacy education for product design engineering students.* Paper presented at Creating Knowledge IV, Copenhagen, Denmark. Retrieved from http://www.ck-iv.dk/papers/PilerotHiort%20Design%20for%20information%20literacy.pdf

Sapp Nelson, M. (2009). Teaching interview skills to undergraduate engineers: An emerging area of library instruction. *Issues in Science & Technology Librarianship, 58.* http://dx.doi.org/10.5062/F4ZK5DMK

PART II

Designing Information-Rich Engineering Design Experiences

CHAPTER **5**

ACT ETHICALLY

Design with Integrity

Megan Sapp Nelson, Purdue University

Donna Ferullo, Purdue University

Bonnie Osif, The Pennsylvania State University

Learning Objectives

So that you can guide student design teams to identify ethical and social aspects of engineering design, upon reading this chapter you should be able to

- Define and articulate professional integrity as it applies to engineering design

- Identify and apply a code of ethics perspective of professional behavior to an engineering design team project

- Coach students in the ethical use of information throughout the design process

INTRODUCTION

Even before starting a design project, while still organizing the team, instructors frequently begin by setting expectations for student work, including introducing the concepts of ethical behavior. Among other topics, ethical behavior includes doing due diligence, presenting all of the relevant information and not just convenient facts, and respecting the work of others. Ultimately, the goal for engineers is to provide an accurate assessment of the strengths and weaknesses of their solutions, rather than misrepresenting a solution in order to win a contract. Instilling this ethos into the classroom environment from the beginning will create an appropriate focus on engineering design as a knowledge-building activity. It will also reinforce professional skills required by ABET, the accrediting body for engineering programs (student outcome 3) (ABET, 2013).

As students move through their academic career with the goal of becoming a professional engineer, a major outcome is their acculturation into the discipline. One pillar of engineering is professional integrity. The National Society of Professional Engineers (NSPE) mandates in its *Code of Ethics for Engineers* that engineers will "conduct themselves honorably, responsibly, ethically, and lawfully so as to enhance the honor, reputation, and usefulness of the profession" (2007, I.6). Each of these facets grows out of the idea of personal integrity as generally understood in many cultures. While these concepts are prevalent in the dominant culture, how do students learn to recognize situations that require recognition of ethical gray areas, comparing and deciding the relative priorities of competing stakeholders or specifications? The challenge of introducing professional integrity and related concepts of social responsibility, information ethics, and technical competency

is to introduce them within the context of the engineering design process described.

An engineering code of ethics addresses the reality that the work of engineers and the decisions they make have serious implications for a number of people. Unlike a physician or other professional with whom members of the public interact directly, most people do not know the engineer who designed the product they use, the appliance they turn on, or the bridge they drive across. There is an implied social contract that the engineer will act ethically and with integrity. This chapter addresses concepts and techniques for introducing reflection on professional integrity in the context of the engineering design curriculum.

COMMON CHALLENGES FOR STUDENTS

Undergraduate design team members generally lack a perspective that enables them to place their work in broad context with respect to users. In fact, undergraduates have been acculturated by an educational system to believe that the work they do and the things they create in courses have no value beyond their final grade in the class. For an undergraduate design team, considering the ethical implications of the project first requires a major leap in conceptualization on the part of the students that the work they produce has long-lasting implications and impact on others.

Additionally, undergraduates in their late teens and early 20s have not yet fully developed the portions of the brain responsible for ethical reasoning. The prefrontal cortex continues to develop well into the 20s (Sowell, Thompson, Holmes, Jernigan, & Toga, 1999). This area of the brain controls higher order logic, including ethical reasoning (Fumagalli & Priori, 2012). The implications of this physiological fact for undergraduate design teams are that

- students on the teams will have different levels of facility with ethical reasoning;
- ethical reasoning must be deliberately introduced into the pedagogy and conversation of the student design team in a facilitated way in order to assure that ethical implications are considered during the design process;
- ethical constraints that are obvious to the instructor are typically not obvious to their students.

For all of these reasons, ethical reasoning is an aspect of engineering design that can and does cause difficulties for design students.

Undergraduates deepen their appreciation of their personal integrity as they perceive themselves as an adult who controls their own behavior and responses to situations. Developing positions based on reason and evidence, weighing pros and cons, debating differences with peers, and reflecting on the ethics of decision making processes encourages students and helps them to effectively handle ethical quandaries. Education in the area of ethical reasoning assists in the development of students who are ssocially responsible and ethically grounded professional designers upon graduation. As we will see in the next section, engineers are expected to be both.

PROFESSIONAL EXPECTATIONS OF ETHICS AND INTEGRITY

Oakes, Leone, and Gunn (2012) stated that "in addition to technical expertise and professionalism, engineers are also expected by society and by their profession to maintain high standards of ethical conduct in their professional lives" (p. 395). Each profession has its own code of ethics that addresses its uniqueness. Within engineering, many organizations have produced codes of ethics intended to guide decision making and behaviors of professional engineers. A code of ethics for engineers is one with far reaching implications, as the results of engineering design can affect not only the bottom line of a company but actual structures, products, and the lives and safety of those who come in contact with the products of the engineers. Engineering decisions must not be made haphazardly, or be based on personal preference and self-interest. Rather, engineering decisions must be guided by a professional code of ethics, as an overarching set of principles; engineering thinking and judgment, supported by data and analysis and informed by collective knowledge; and wisdom embodied in such things as specifications, standards, codes, and regulations.

"Primarily, a code of ethics provides a framework for ethical judgment for a professional" (Fleddermann, 2012, p. 25). There are a number of codes of ethics for engineers. Most professional associations have their own codes, and this can range from a few lines to the several-page-long detailed list of the NSPE. The importance of ethics to the profession is made clear by the inclusion of codes of ethics on all major engineering society Web pages and in the Criteria for Accrediting Engineering Programs from the ABET (2013).

In general, all of the codes have a statement supporting engineering for public safety, honesty, and integrity in design. They generally agree that engineers are to put society first and design only in areas of competency, call for objectivity and truthfulness in disclosures and

> **BOX 5.1**
> **Code of Ethics Websites**
> *American Society of Civil Engineers (ASCE)*
>   "Code of Ethics": http://www.asce.org/Leadership-and-Management/Ethics/Code-of-Ethics
> *ASME Standards Technology, LLC*
>   "Ethics": http://files.asme.org/STLLC/13093.pdf
> *Institution of Civil Engineers (ICE)*
>   "Code of professional conduct for members": http://www.ice.org.uk/Information-resources/
>   Document-Library/Code-of-professional-conduct-for-members
> *National Society of Professional Engineers (NSPE)*
>   "Ethics": http://www.nspe.org/Ethics/CodeofEthics/index.html
> *Online Ethics Center for Engineering and Research*
>   http://www.onlineethics.org
> *Institute of Electrical and Electronics Engineers (IEEE)*
>   "IEEE Code of Ethics": http://www.ieee.org/about/corporate/governance/p7-8.html

dealings, and focus on the personal integrity of all engineers.

A code of ethics is a starting point, but it cannot be considered comprehensive as there are specifics and situations that cannot be addressed directly by the principles of the code. But, "a code expresses these principles in a coherent, comprehensive and accessible manner. Finally, a code defines the roles and responsibilities of professionals" (Fleddermann, 2012, p. 25). A representative list of current code of ethics websites is contained in Box 5.1.

THE CONCEPT OF PROFESSIONAL INTEGRITY

The extent to which individuals in our complex technological society can control the risks that they are exposed to is severely limited. . . . There is no practical way for each of us (even as engineers or scientists) to evaluate the degrees of safety designed into the many consumer products that we use. . . . It is thus of great importance that engineers recognize their professional responsibilities with respect to human safety, that they be properly educated to fulfill those responsibilities, and that they be given adequate authority to carry them out. (Unger, 1982, p. 12)

As discussed, integrity is a crucial aspect of the job for a professional engineer. As defined by the NSPE, honor, ethics, responsibility, and lawfulness are the most fundamental behaviors to be displayed by engineers (see Box 5.2). Only if these traits are present in conjunction with disciplinary knowledge and technical skills is a person a fully qualified engineer. Engineering has been characterized as being "essential to our health, happiness and safety" as "engineers help shape the future" (National Academy of Engineering, 2008, p. 8). In doing so, engineering as a discipline explicitly seeks to act in an ethical manner in relation to the stakeholders (and increasingly, environment) it serves.

Undergraduate engineering students may consider social responsibility either an obvious

---

BOX 5.2
NSPE and ASCE Codes of Ethics

National Society of Professional Engineers **Code of Ethics for Engineers** ***Fundamental Canons***[1]
Engineers, in the fulfillment of their professional duties, shall:
1. Hold paramount the safety, health, and welfare of the public.
2. Perform services only in areas of their competence.
3. Issue public statements only in an objective and truthful manner.
4. Act for each employer or client as faithful agents or trustees.
5. Avoid deceptive acts.
6. Conduct themselves honorably, responsibly, ethically, and lawfully so as to enhance the honor, reputation, and usefulness of the profession.

American Society for Civil Engineering **Code of Ethics** ***Fundamental Canons***[2]
1. Engineers shall hold paramount the safety, health and welfare of the public and shall strive to comply with the principles of sustainable development in the performance of their professional duties.
2. Engineers shall perform services only in areas of their competence.
3. Engineers shall issue public statements only in an objective and truthful manner.
4. Engineers shall act in professional matters for each employer or client as faithful agents or trustees, and shall avoid conflicts of interest.
5. Engineers shall build their professional reputation on the merit of their services and shall not compete unfairly with others.
6. Engineers shall act in such a manner as to uphold and enhance the honor, integrity, and dignity of the engineering profession and shall act with zero-tolerance for bribery, fraud, and corruption.
7. Engineers shall continue their professional development throughout their careers, and shall provide opportunities for the professional development of those engineers under their supervision.

[1] PDF available for download at http://www.nspe.org/Ethics/CodeofEthics/index.html.
[2] PDF available for download at http://www.asce.org/Leadership-and-Management/Ethics/Code-of-Ethics.

---

and commonsense fundamental or a nonobvious and unduly complicating aspect of the design process—perhaps even not part of their engineering design considerations. As noted above, which view a student takes may have much to do with the state of development of his or her brain and reasoning abilities. Nevertheless, all students can be taught to consider the function of social responsibility in engineering design and its implications for their specific project.

Social responsibility includes considerations of the diverse range of individuals who may interact with the artifact they design. The most common consideration is the impact on stakeholders, whether the direct client or downstream users (see Chapter 7 for more information regarding user groups). However, social responsibility also includes considerations of environmental impact and sustainability, and legal and regulatory responsibilities (including intellectual property).

Sustainability is at essence represented by the three Ps (Jonker & Harmsen, 2012, p. 10):

- "People", [*sic*] the social consequences of its actions
- "Planet", the ecological consequences
- "Profit", the economic profitability of companies (being the source of "Prosperity")

Within the design context, sustainability requires that the artifact honors the integrity of the stakeholders, the current environment, and the business bottom line.

Engineering is a global discipline. A product, system, or process designed in the United States may be manufactured in Southeast Asia with raw materials mined and shipped from Africa, Russia, and the Middle East and be packaged and shipped back to the United States for sale in a retail establishment. The situation of a design artifact is most likely much more global than undergraduate design students may realize (Luegenbiehl, 2010).

Development of a situational awareness that fully anticipates the impact of a design project is a part of developing a sustainable artifact. "Sustainability can be approached from many different perspectives, varying from North to South throughout the world, and from governmental regulations to market considerations" (Jonker & Harmsen, 2012, p. 2).

An important part of designing for sustainability is learning from all invested parties and creating the best possible solution to meet their needs and expectations. In the context of a student design team, many different perspectives can be facilitated by encouraging all voices on the team, including those with non-majority backgrounds, to contribute. Students of diverse and international backgrounds bring different insights and assets to the design process. Often these participants in the design must be encouraged to share their strengths in group interactions. Majority students in a design team frequently have a difficult time recognizing the value in the variety of experiences on the team, as they rush to a design solution that frequently arises out of the input of the most assertive team members. Eliciting valuable experience and input from non-majority team members

is similar to eliciting design constraints (discussed in Chapter 7).

COMPETENCY

A major facet of engineering ethics is simply to acknowledge what you don't know, when you don't know it. Most codes of engineering ethics require that engineers not perform work or give advice beyond the limits of their technical knowledge and competence. Competent engineers honestly assess their own ability to complete a project well and on time. By extension, engineers will

- refuse to sign documents that they do not understand;
- identify projects that are not up to relevant codes or standards as well as refuse to sign documents for those projects;
- seek out experts to complete work that they feel is outside their personal competency.

Students are working to achieve competency in engineering and engineering design, while also attempting to develop an internal gauge for what skills they possess. Tools such as skills assessments, completed by the individual or team, provide insight into the skills in which students and their teammates appear strongest and weakest. Skills inventories and assessments can be utilized throughout the design process with a variety of outcomes. A baseline can be established early in the semester using a skills assessment. With the addition of a post-assessment, changes in perceived skills can be measured.

Building recognition of personal competence can be woven throughout the design process. As part of early team-building exercises, students can develop individual so-called elevator pitches that describe their areas of special knowledge, skills, and competence. Teams can then create a team consulting brochure intended to give the stakeholders an understanding of the expertise represented.

Student teams can be prompted to develop procedures for the distribution of tasks based on either strengths (for quick turnaround on a deliverable) or weaknesses (to build competency across the entire team). They can also work together to identify competency gaps across the team and invite an expert to fill in that weakness. Students should be encouraged to identify alternative solutions that may reduce the need for a weaker skill and to determine the difference in resources (time, money, effort, physical materials) required by both strengths-based and competency-building task distribution.

Instructors should take the opportunity to help students grapple with the concept that an engineer cannot be excellent at all aspects of engineering. As such, students should be prepared to network with other experts upon their graduation to build up their informal ties in preparation for future project needs. By building this capacity for networking throughout the undergraduate engineering curriculum, students are investing in lifelong learning habits that will enable them to identify, articulate, and track their expanding professional competencies. Most students will not make this mental connection between their own skills inventory and networking unless an instructor invests time in introducing them to that concept.

OBJECTIVITY

Objectivity is the active pursuit of presenting the complete context of design decisions and constraints in a manner that is absent of bias, prejudice, and emotional influence. There are a number of concepts underpinning this definition, as listed below.

- Engineers choose to be objective. Action must be taken to increase objectivity; it does not naturally occur. Engineers, along with everyone else in human society, are prone to prejudices and biases, often unknown to themselves. In order to be truly objective, an individual must choose to set aside his or her own personal inclinations.
- Objectivity has as its goal the removal of the engineer's personal prejudices and biases from an engineering decision. Therefore, engineers seek to present the full context of how decisions are made in order to allow stakeholders to develop their own opinions.
- Objectivity is an external discipline, as opposed to an internal state. Engineers will of course have opinions of their own about specific aspects of a deliverable. Objectivity ensures that the stakeholder has the full information necessary to make decisions without exposure to prejudice.
- Objectivity is a mitigating technique for separating the engineer as an individual from the product he or she has created.

In the engineering design classroom, objectivity can be practiced in a number of ways. While case studies are frequently used to discuss issues of objectivity as well as other ethical canons, students learn best through active engagement and practice.

Early in the design process, while gathering design constraints from stakeholders, students can begin to examine their own biases and prejudices through reflective exercises such as journaling. Identifying preconceived notions or preexisting biases will help students to mitigate their impact on the design product.

After the design constraints have been gathered and specifications developed, students have the opportunity to systematically check their implicit assumptions by presenting a document for the approval of the stakeholders that details the constraints found during specification development and how the design specifications mitigate those constraints. This allows the teams to check their own understanding of the design context, while confirming that the stakeholders feel that the ultimate deliverable will meet the constraints. It also allows the students to present information in an objective way, neither pushing the stakeholder to accept the specifications as written, nor influencing the stakeholders' decision.

The development of documentation also provides an opportunity for practicing objectivity. Requiring students to include critical assessments of the resources they are using to assist in the conceptual design, detailed design, and fabrication stages of the project not only creates an extensive paper trail for why decisions were made throughout the project, but also allows students to practice evaluating what sources of information should be shared with the stakeholders.

TRUTHFULNESS

Another important aspect of information ethics that is required of both professional engineers and engineering design students is truthfulness. Truthfulness is the avoidance of deceit, whether through commission or omission of communicating relevant information. In engineering, truthfulness is paired with objectivity to create a situation in which full disclosure is made to a stakeholder or in another business relationship. Honesty is particularly key to

the decision-making process; in the absence of a truthful disclosure, major flaws in a design product or process are not identifiable because the full context has been hidden or altered. Stakeholders rely on engineers to provide truthful information.

A major component of this, providing full access to all relevant and pertinent information, is similar to objectivity. For undergraduate students, the ability to identify relevant and pertinent information is a skill that needs to be introduced. While students may have written term papers previously in their academic career, they commonly have not yet realized that the same information retrieval, synthesis, and citation skills are relevant to their engineering projects. Requiring citation of all sources of information used to create documentation goes a long way toward improving the quality of undergraduate project documentation, while simultaneously helping the students remember the importance of truthfulness. (See Chapters 6 and 13 for more information on communicating via documentation.)

Attribution and acknowledgment are an equally important part of being a truthful engineer. Acknowledging the work that someone else has done to create the artifact is both ethical and courteous. Work that has been taken without attribution is plagiarized. Plagiarism could end an engineering career in academia and may hurt the professional reputation of an engineer for many years.

Attribution and acknowledgment are connected with the competence of the engineer. No one engineer has the expertise to complete a large project by him- or herself, and many small projects are also team based. Recognition of the expertise of everyone who participated elevates the perceived competence of the resulting product because the competence of the team is broader and deeper than that of one individual alone.

Generally, students understand the concept of truthfulness through the lens of their own cultural background. Raising awareness of truthfulness during the process of an engineering design class simply requires that accountability be built into the system. One way is to require students to cite the resources that they are using to develop the design product. Sources of information are not uniformly of high quality. This exercise allows the instructor to help students understand that the credibility of the sources they have chosen reflects on their credibility as a competent engineer.

Students may also keep design notebooks. If so, the design notebooks should be graded in such a way that the contribution of individual members of the team are placed within the context of decisions the whole team is making. In that way students are able to identify who contributed what to the team and also to understand their role within the work of the group, thereby identifying growing competencies for themselves. A related opportunity comes in the form of individual portfolios of work, which some schools are now requiring for their undergraduate students. Helping the students to identify specific areas of expertise within a project and then truthfully place their work within the larger scope of the team's design process will assist students to identify their own competencies, which will ultimately impress employers.

CONFIDENTIALITY

While objective and truthful disclosure is valued for engineers, in addition information can be very valuable and therefore must be controlled in the timing and breadth of the disclosure. Many engineers are asked to sign confidentiality or nondisclosure agreements to work on a particular project. These agreements limit whom can be

told details about the project, or even whether the project exists. A nondisclosure agreement generally contains language specifying what information is within the scope of the agreement, permissible ways for the information to be used, and how or when the agreement will end.

Engineers agree to be truthful and honorable by seeking to abide by codes of ethics. As such, if an engineer has signed a nondisclosure agreement, he or she is bound by the terms of that document. Therefore, each document signed needs to be carefully read and understood, questions should be asked if any part of the document is difficult to understand or abide by, and the document should be examined for requirements that raise professional and personal ethical questions that would make it difficult for the engineer to abide by the agreement. These red flag issues should be discussed.

To assist undergraduates in their future career, discussing the contents of a nondisclosure agreement within the context of a design assignment is appropriate. In some cases in which corporations are the clients for a project-based learning class, the students may have a legally binding nondisclosure agreement that they must sign before beginning the project. Breaking down a real or sample agreement, encourages students to identify the governing terms of a nondisclosure agreement, identify potential terms that would be likely sources of noncompliance, and discuss what they are agreeing to abide by.

INTELLECTUAL PROPERTY

As members of a design team, the students are creating something original, perhaps for the first time in their career. As such, they are working as engineers with a vested interest in intellectual property. To act as honorable and

---

BOX 5.3

Definitions of Intellectual Property Terms

Copyright—Federal law that protects creative works that are unique in some manner and that have been expressed in a tangible form. Copyright protects a whole cadre of works such as books, journals, music, computer programs, and images. Ideas are not protected under copyright law. It is the expression of the idea that generates the protection. Procedures, processes, systems, and methods are not copyrightable. (See patents). Copyright limits the amount of time the copyright holder can retain the rights to the work.

Trademark—A distinctive name, slogan, symbol, or design that identifies and distinguishes the product or service from other brands. Example: Nike as the name as well as the swoosh mark. Trademarks protect a trade or a service.

Trade secret/trade dress—Similar to trademark. Trade secret protects vital processes or components of a product. Trade dress protects the overall appearance of a design. Example: Coca Cola's recipe is a trade secret. The distinctive red and white packaging is trade dress.

Patent—Legal document claiming ownership of a unique function (utility patent), hybridization (plant patent), or aesthetic (design patent). A utility patent can be classified as a machine, a process, a composition, an article of manufacture, composition of matter, or any new and useful improvement to an invention.

Prior art—Preexisting information describing a process, product, procedure, system, or method for the patent process.

Right of publicity—The control of the commercial use of an individual's name, image, and likeness that can continue even after death.

---

responsible engineering designers, students need both to acknowledge the influence that preexisting artifacts have had on their product, as well as to identify the work for which individual students are responsible. Acknowledging the work of others creates transparency and exemplifies the honesty of the engineer doing the work. Similarly, by identifying those portions of the work that the engineering student created, the student is taking responsibility for the quality and completion of the work.

Intellectual property is a highly visible, strongly codified aspect of legal and ethical behavior associated with design and is made up of a number of legal frameworks that protect the work that has been done. For most enterprises, it is a financial imperative to protect intellectual property; frequently it is the core asset owned by a company. The world of intellectual property revolves around the common theme of protecting intellectual output, which can be manifested in many forms and in many ways. The existence of a nondisclosure/confidentiality agreement generally signals a belief that the project that is being completed is a potential source of disclosure of existing intellectual property and development of new intellectual property. This document seeks to protect intellectual property.

Intellectual property is a possession similar to real property such as homes and cars in that there are laws that protect and sometimes dictate its ownership. Intellectual property violations are identifiable via design documents and the final product, while simultaneously enforceable in courts of law.

The area of intellectual property law consists of copyright, trademark, trade secret/trade dress, patents, and right of publicity. Each of these areas has its own unique protections (see Box 5.3).

To assist students to develop their knowledge of intellectual property and how it works in context, it is recommended that consideration of each intellectual property concept be intentionally included appropriately into the design cycle. Many of these are directly or indirectly utilized by students in the process as it is. Design artifacts and notebooks, manifestations of the engineering design decisions made, are the physical proof of reasoned ethical decision making.

COPYRIGHT

Copyright comes into play during the specification and conceptual design phases of the design cycle. Students will be accessing a number of information resources, nearly all of which will be governed by copyright or an alternative intellectual property agreement such as open source or Creative Commons licensing (see Box 5.4). Copyrighted works are protected even if they are freely accessible or given away, whether print, electronic, or digital media.

In the educational setting, engineers have the option of fair use at their disposal which allows specific uses of copyrighted informa-

tion. Whether fair use applies is determined partially by whether the information is used in an educational setting, how much of the work is copied, how unique the original work is (fiction is protected more heavily than factually based work), and the financial impact on the market for the original work. Each of these factors has implication for engineering design.

While students are attending a university, much of the information that they are using is governed by the educational exception to copyright, meaning that the expectation of paying revenues for use of the work is significantly lower than if they are professional engineers who are using information for commercial use. Using a small percentage of a given work (a sentence, a paragraph) is considered to be considerably fairer than using entire chapters or whole works without permission. Many e-book providers limit the amount that can be downloaded from any one work for this reason. Generally in engineering, the information used is factually based, which means that the usage terms may be more lenient. The possible negative financial implications from the use of a copyrighted work are particularly relevant to digital media. If artwork or images are used in the creation of a deliverable but copyright is not honored, artists will lose money for work

---

BOX 5.4

Open Source and Creative Commons Licensing Websites

Explore these websites for more information on open source and Creative Commons licensing:

- http://creativecommons.org/about
- http://orbison.exp.sis.pitt.edu:8080/webdav/Miscellaneous/understanding-common-open-source-licenses.pdf
- http://opensource.org/licenses
- http://www.gnu.org/licenses/license-list.html

that they distributed for the purpose of making money.

As future engineers, it is important for students to recognize that the work that has been distributed, whether via the Web or in print, has economic value. As creators of information, the honorable as well as legally required course of action is to comply with copyright when appropriate. If an exception such as fair use does not apply, then it is the responsibility of the user of the copyrighted work to seek permission from the copyright holder. A copyright infringement of a work transpires when the use made of the work is outside of the exceptions such as fair use and/or permission was not granted.

PATENTS

Patents are generally accessed during the specification and conceptual design phases, although they may also be used during detailed design. Patents protect the intellectual property rights of an inventor or patent holder and ensure that the patent holder has time to commercialize the invention before competition can produce the product as well. As part of the process of determining prior art, students should be looking for patents that currently exist. As part of a truthful, objective, and comprehensive background search, patents should be included. If the project is one that is novel enough to be

For engineers who also have an interest in law and a detail-oriented mindset, the profession of patent attorney can be lucrative. Students generally need to hold a bachelor's degree in a scientific field, then attend law school, earn a JD, and pass their state bar examination. Patent attorneys can work for the U.S. Patent and Trademark Office or in private practice.

commercialized, the failure to conduct a prior art search may lead to the product's failure due to patent infringement. It also casts doubt on the credibility of the engineering team who designed the product.

If a patent search is assigned, students should be encouraged to consult with a local librarian. The dictionary of terminology used to describe patents is quite different from the everyday terminology that society uses to describe those items. What is known as a "generally spherical object with floppy filaments to promote sure capture" in the patent database is known as the Koosh ball in general society.

Librarians can help to increase the success of beginner patent searchers by providing coaching on the selection of terminology for keyword searching and classification searching (which enables the searcher to find a number of related examples at once as opposed to an individual patent). The entire U.S. Patent Database back to its inception in 1790 is available via the uspto.gov website.

SUMMARY

Engineering students must have a well-developed sense of professional integrity. This will manifest itself in their student group work and professional lives through evidence of the consideration of the safety, health, and welfare of others, through the development of competency and the restriction of work only to those areas of competency, and through a robust understanding of information ethics. Student design projects present a high-impact teachable moment—an opportunity for students to practice ethical reasoning and develop both a stronger sense of self and responsibility to stakeholders. Beginning the discussion of ethics and setting expectations for individual and team ethical behavior, including ethical use of information, at the outset of a project when teams are formed, provides a foundation that will serve students well not only in their course work but also in their careers after graduation.

SELECTED EXERCISES

Exercise 5.1

Using engineering controversies as a conversation starter for a class discussion, followed by an individual reflection activity, can provide a baseline at the beginning of the semester to understand the relative ethical reasoning abilities of the students in a class. The same topics can be used to start required blog or wiki conversations. Some possible topics include the following:

- MIT/Aaron Schwartz case of downloading scholarly articles illegally
- Algo Centre Mall roof collapse
- URS Corporation and the Minneapolis I-35 W bridge collapse
- Sinking of the Titanic
- Bhopal chemical disaster
- Chernobyl nuclear power disaster
- Fukushima nuclear power disaster
- Charles de Gaulle Airport roof collapse
- Banqiao Dam disaster
- Niger Delta contamination

For more information on potential questions to pose and ideas for other case studies, see the Online Ethics Center website, http://www.onlineethics.org.

Exercise 5.2

A service learning class is partnered with a nongovernmental organization in a Sub-Saharan African country. The students will be partnered with the NGO (nongovernmental organization) staff, who will be the primary interface on the ground between the stakeholder community and the class. The students are tasked with designing a water filter using locally available, sustainable, and renewable sources. A first activity that would enhance objectivity is having them list the assumptions they have about the community, the environment, the stakeholders, and the long-distance communication process. The instructor may require the students to submit their responses and reply back privately while correcting major potential biases and prejudices. The instructor may also initiate a group discussion on the most prevalent assumptions in the class regarding these aspects of the design constraints. Either way, identifying these assumptions early will help the class to avoid the pitfalls of prejudice and bias from the start of the project.

REFERENCES

ABET. (2013). *Criteria for accrediting engineering programs 2012–2013.* Baltimore: ABET. Retrieved from http://www.abet.org/DisplayTemplates/DocsHandbook.aspx?id=3143

Fleddermann, C. B. (2011). *Engineering ethics* (4th ed.). Upper Saddle River, NJ: Prentice Hall.

Fumagalli, M., & Priori, A. (2012). Functional and clinical neuroanatomy of morality. *Brain*, *135*(7), 2006–2021. http://dx.doi.org/10.1093/brain/awr334

Jonker, G., & Harmsen, J. (2012). *Engineering for sustainability: A practical guide for sustainable design.* Amsterdam, The Netherlands: Elsevier B. V. http://dx.doi.org/10.1016/B978-0-444-53846-8.01001-5

Luegenbiehl, H. C. (2010). Ethical principles for engineers in a global environment. *Philosophy of Engineering and Technology, 2,* 147–159. http://dx.doi.org/10.1007/978-90-481-2804-4_13

National Academy of Engineering. (2008). *Changing the conversation: Messages for improving public understanding of engineering.* Washington, DC: The National Academies Press. Retrieved from http://www.nae.edu/19582/reports/24985.aspx

National Society of Professional Engineers. (2007). *NSPE code of ethics for engineers.* Alexandria, VA: Author. Retrieved from http://www.nspe.org/Ethics/CodeofEthics/index.html

Oakes, W. C., Leone, L. L., & Gunn, C. J. (2012). *Engineering your future: A comprehensive introduction to engineering* (7th ed.). New York: Oxford.

Sowell, E., Thompson, P., Holmes, C., Jernigan, T., & Toga, A. (1999). In vivo evidence for post-adolescent brain maturation in frontal and striatal regions. *Nature Neuroscience, 2*(10), 859–861.

Unger, S. H. (1982). Role of engineering schools in promoting design safety. *IEEE Technology and Society Magazine, 1*(4), 9–12. http://dx.doi.org/10.1109/MTAS.1982.5009730

Organize Your Team

CHAPTER **6**

BUILD A FIRM FOUNDATION

*Managing Project Knowledge
Efficiently and Effectively*

Jon Jeffryes, University of Minnesota

Learning Objectives

*So that you can guide student design teams on effective
strategies to plan and manage information and knowledge
collection critical to their project, upon reading this chapter
you should be able to*

- Describe the major information literacy concepts
  critical to successful knowledge management in a
  student team design project

- Identify common problems student teams have in
  developing, implementing, and maintaining an effective
  and efficient knowledge management plan and
  strategies to overcome these

- Describe the pros and cons of various computer-based
  tools, including citation management systems, to use as
  part of a successful knowledge management plan

INTRODUCTION

Before giving a design brief to student teams, instructors generally have them engage in some team organization activities, such as determining roles and developing a shared understanding of responsibility and accountability. One of the organizing activities frequently neglected, however, is determining how students will manage the information they gather and the knowledge they generate so that the whole team benefits. If they do discuss it, students may only go as far as saying they will set up a shared folder on Dropbox or Google Drive to hold their work. However, even if students have thought about a platform, they typically haven't thought about a process for organizing or communicating new information on that platform. Just as piling heaps of papers on one's desk doesn't constitute an effective organizing solution, especially for others trying to find a particular paper in one's filing system, dumping files into a shared folder likewise can lead to much confusion and inefficiency for the team.

Managing information and team knowledge are keys to the success of any design project. In 1986, the world witnessed one of the most dramatic and tragic design failures in modern history when the space shuttle *Challenger* exploded shortly after takeoff, killing all seven of its crew members. After a lengthy review, investigators found that the tragedy did not stem from a lack of information or bad data, but rather "failures in communication . . . based on incomplete and sometimes misleading information" (Presidential Commission on the Space Shuttle Challenger Accident, 1986).

As the *Challenger* explosion showed only too tragically, a well thought out plan for storing and communicating the information that each team member accrues during the course of a design project is necessary for a successful team project. This extends to the new knowledge generated by the team during the course of their project. As well as helping to avoid design failures, a thorough knowledge management plan can expedite the work of the team, making it more efficient and effective, and save time for all team members throughout the design process.

Knowledge management can most succinctly be defined as "the management of knowledge workers as well as the information they deal with" (Statt, 2004, p. 81). Kraaijenbrink and Wijnhoven (2006) expand that description, stating that "as an academic field, knowledge management has concentrated on the creation, storage, retrieval, transfer, and applications of knowledge within organizations" (p. 180). The literature on knowledge management explores further complexities (see Bredillet, 2004, for a nice introduction), but for the purposes of this chapter we will explore the topic using these more practical definitions focusing on the way information is managed throughout an organization, in this case an engineering student design team.

COMMON CHALLENGES FOR STUDENTS

The most difficult challenges design teams encounter in setting up a robust information management plan are motivation and time. Sitting down to have a conversation about how to share information and exchange knowledge is probably the least exciting part of a design project. Students will be keen to jump right into their first opportunities to practically apply all the technical skills they've been amassing during their college experience without considering future issues such as information management. Also, to make a thorough plan will take a considerable amount of time. For students with a full slate of classes and other activities, making the time up front to formulate a plan tends to be a lower priority (even with the promise of long-term time savings). To ensure the inclusion of this step, modeling sound design practice, the instructor should include it as the focus of a classroom session and make a formal, well-documented plan a graded deliverable of the project. To guarantee that students take the time to comply with the plan throughout the design process, each design team should designate a member with the responsibility of monitoring the information sharing in the role of an information manager.

INFORMATION LITERACY AND KNOWLEDGE MANAGEMENT

In their discussion of knowledge management, Kraaijenbrink and Wijnhoven (2006) describe a process of knowledge integration, made up "of three stages—identification, acquisition, and utilization of external knowledge" (p. 180). This process makes the most sense for the integration of information literacy skills. Returning to the facets of information literacy outlined in Chapter 2, this process maps nicely to the facets of locating information and evaluating information. Using Kuhlthau's (2004) Information Search Process, this step of the engineering design process would fall under the collection stage.

As outlined in Chapter 4, the introduction of information management occurs early

in the Information-Rich Engineering Design (I-RED) model as the activity "organize the team." Introduction of these concepts at the beginning of the design process will prepare the team for success. This foundational skill sets the direction for the entire design project and needs to be addressed throughout the design process and over the design iterations. Engineering librarians will focus instructional efforts on the organization and communication of information gathered during the design process in literature reviews, collection of prior art, and searches for relevant standards and regulations that may impact the engineering design. The instructors can then correlate these practices to other steps such as experimental data management and collecting stakeholder feedback.

The connection between information literacy and knowledge management has been examined by Singh (2008), who found that "IL [information literacy] facilitates sense-making and reduction of vast quantities of information into fundamental patterns into a given context. That is also the heart of the matter in knowledge management" (p. 14).

O'Sullivan (2002) also examined the connection between information literacy and knowledge management and found that even when the corporate world does not use the terminology employed by their library counterparts, they do value the skill set required by both information literacy and knowledge management as integral to success in the workplace. Singh (2008) reinforces the importance of information literacy, placing it at the foundation of knowledge management. Engineering students may not engage intentionally with information literacy at this stage of their engineering design experience, but often the skills they are beginning to employ fall into this skill set. The engineering librarian can bring more

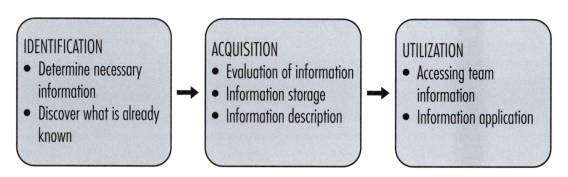

FIGURE 6.1   Information literacy within knowledge integration. (Modified from Kraaijenbrink & Wijnhoven, 2006.)

explicit understanding of these skills and their benefits into this early portion of the design process, setting the foundation for an information-enriched design process.

INTEGRATING INFORMATION LITERACY

Figure 6.1 incorporates information literacy into Kraaijenbrink and Wijnhoven's (2005) conception of knowledge integration.

For the purposes of an engineering design project this process is linear, but it will repeat throughout the design as students enter different stages of their project. Students follow the process outlined in Figure 6.1 while conducting their search for existing information in a literature review, and then start the process over when they start generating their own information in the experimental stage.

To establish good practices, a session on information management should occur early in the design project and focus on how the team plans to manage and communicate the process listed in Figure 6.1. Since the early stages of design include identifying relevant information that already exists, the focus of the illustration uses a literature search as its example. Cita-

tion management software provides a means of managing the information acquired during this stage of the design process.

CITATION MANAGEMENT

Citation management software provides an intuitive point of entry to integrate information literacy skills into the information management portion of engineering design. The software allows students to collaborate in the collection and organization of citations and subsequently output those citations into formatted bibliographies and in-text citations (see Box 6.1).

Childress (2011) has previously discussed the role of citation management software in library instruction. This software often falls in engineering librarians' wheelhouse because of their expertise in using scholarly citations, or because the library finances access to the tool(s). Librarians can exploit their mastery of these tools to simultaneously insert information literacy skills into the early stages of a design class and lay the foundation for the use of best practices in information management throughout the engineering design process.

Students easily recognize the value of citation management software for their course work and work flows. It can save students time

---

BOX 6.1

Citation Management Tools

EndNote

Fee-based citation management software. Downloads directly to the user's hard drive. Syncing and collaboration are available through EndNote Web.

Mendeley

Basic edition is free to download to the user's hard drive. Allows for online syncing and collaboration with groups. Basic edition limits number of groups as well as number of collaborators.

RefWorks

Fee-based citation management software that is entirely cloud based. With institutional subscription, students can have multiple accounts, allowing design teams to create a shared account.

Zotero

Free download is available online. Can be installed as Firefox plug-in or as a standalone program on the user's hard drive. Allows for online syncing and collaboration with groups at no additional cost.

---

and prevent instructors from puzzling through incomplete or poorly formatted citations. These time-saving aspects capture a classroom's attention and open the door for receptivity to information literacy skills. Duong (2010) has written specifically of the value of science librarians using Zotero in an outreach effort.

Citation management software can be divided into two major forms: fee-based and freeware. The fee-based citation managers (such as RefWorks and EndNote) are only available through institutional site licenses or personal purchases. Freeware programs (such as Zotero and Mendeley) provide a free basic software package and then charge for added functionality, such as extended cloud-based storage space and large group collaboration functionality.

The engineering librarian and design instructor can determine which tool to incorporate into the class, but the evaluation and ultimate decision making can also be incorporated as a piece of the instruction itself—the engineering librarian providing students with the strengths and weaknesses of each tool and letting them critically engage with the information and decide which program will work for their individual group. Regardless of the type of software ultimately selected, most citation managers facilitate collaboration and organization through the creation of groups (sometimes also referred to as folders or libraries depending on the particular software—all the different terms provide the same type of functionality).

IN THE CLASSROOM

Ideally, citation management is introduced as part of an integrated, intentional information gathering process. Instruction starts with an introduction to the knowledge integration process outlined in Figure 6.1 and provides an overview of the different types of literature available and relevant to engineering design, as well as the tools available to locate this information. (More details on the different kinds and purposes of technical literature are covered in the following chapters.) The instructor, often an engineering librarian, provides a short lecture at the beginning of the classroom session, but this instruction might be covered in earlier course work or given as a pre-class video tutorial. The introductory content describes the development of a literature review strategy at the outset of the project and includes an offer of consultative services from the engineering librarian to the group for further, personalized guidance on which information resources might work well for their project.

After students are familiar with the variety of information types available, the instructor introduces the mechanics of the citation management software (in this example, the instructor and engineering librarian choose one citation manager that the entire class will use). This introduction provides a brief, general explanation of the functionality that the software offers and covers the mechanics of importing citations from indexing databases into a collaborative citation management group. The interaction between database and citation management software differs from database to database. This fact, often frustrating to the user, provides the engineering librarian the opportunity to showcase multiple information sources to the students. In discussing the steps necessary to retrieve citations from the article database, the instructor can also point out the differences in the citations that result from searching multiple information resources for articles on the same topic. These demonstrations also illustrate how word choice impacts results—modeling an ideal information-literate process.

An active learning exercise follows this short introduction and demonstration. Students are directed to work in their design teams to create a list of the types of literature they want to explore and the resources they plan to search. They will start to create a literature review plan, assigning individuals to particular resources and setting a deadline for completion. At the end of the discussion each team sets up a citation manager account and practices getting at least one citation into their library. At the end of the exercise, the instructor pulls the class together and connects the work they have just completed to the "Identification" stage of knowledge integration outlined in Figure 6.1.

Now the students have an account started and at least one citation included in their library. The instructor moves the presentation along to the collaborative use of descriptive tags and "Notes" fields of the citation record. These descriptors can be informative (i.e., where the design student located the information) or evaluative (i.e., the relevancy of the article to their project). These features of citation management software foster communication among the group members. The engineering librarian models effective practices—such as creating an article ranking terminology, noting who added or read a citation, and documenting the resource searched and the terms used to find the information—but ultimately the individual design teams determine their own unique methodology to employ these features.

The engineering librarian stresses the importance of agreeing on a standard descriptive practice early in the design process and employing it uniformly throughout the project. Following the routine ensures the most efficient use of student time, reducing the chance of duplication of work for the entire design team. This practice also illustrates the iterative nature of the research process. At the end of the process the design students will see that multiple search terms, employed in various information resources, were necessary for a comprehensive review of the current state of their design topic. These descriptors will also track the iterative nature of the design process itself, providing a record for the different approaches the team takes in regard to their design problem.

As mentioned, the notes and tags feature of the citation management software can also be used in the critical evaluation of information resources. A tagging structure based on the relevance and quality of the information included in the corresponding citation helps the whole design team quickly identify the best resources for their project. It also demonstrates that not all information is created equal and that every resource must be read with a discerning eye. The tagging process also fosters critical dialogue when disagreements arise on the qualitative values noted. The notes feature can also be used to highlight particular portions of an article that are especially relevant to the research project (e.g., "look over pp. 20–22—skip the rest"). Once again the selling point to students will be that they are saving time for their group and increasing their efficiency, but at the same time the librarian advocates a critical engagement with every text and reaffirms that not everyone must read every article from abstract to bibliography.

At this point, the engineering librarian facilitates another learning activity. Students reconvene in their groups and discuss a standard descriptive practice to be used in the information management of their literature review. After the group discussion, students report out to the entire class for comment in order to facilitate peer learning. The instructor connects the work completed in the activity to the development of the "Acquisition" stage of knowledge integration outlined in Figure 6.1.

Following this discussion, the engineering librarian demonstrates the feature of the citation management software that automatically generates formatted bibliographies. This feature often captures the students' attention and demonstrates a concrete benefit that will result from their use of the citation manager. The bibliography-creation functionality can play an important role in the ethical use of information as well as in communicating with stakeholders about the team's progress. The instructor connects the demonstration to the "Utilization" stage of knowledge integration outlined in Figure 6.1.

Along with providing the design groups with efficiency-enhancing tools and introducing (or reinforcing) information literacy con-

cepts, this session also models best practices in communication and transparency of process that should be employed throughout the entire information management process of the design project, including experimental methods, test findings, stakeholder feedback, and so forth. At the end of the session the course instructor brings the students' attention back to the knowledge integration model and discusses how they will want to come up with a standardized plan for managing their information at all stages of their design work. Just as they have developed procedures for sharing their literature resources, students will also need to make an agreed upon method for sharing the information they gather from all the different aspects of their design work. The session demonstrates how open communication and codified standard procedures provide the most efficient experience in team-based design work.

EVALUATION OF INTERVENTIONS

The active learning session outlined in the previous section provides multiple opportunities for the instructor to check in and provide formative assessment to ensure that students understand the content covered in the classroom session. As an assignment following this class session, students should be asked to submit a formal information management strategy for review as a deliverable of their project. In reviewing the plan the instructor and librarian will want to ensure that this strategy includes all three steps of the knowledge integration outlined previously. A rubric of all the details the instructor would like to see in the finalized plan (see Table 6.1) will help with consistent evaluation. If key components are missing, the instructor or librarian can provide point-of-need

TABLE 6.1 *Example Assessment Rubric for Knowledge Management Plan*

Criteria	Level of Achievement		
	Poor	**Satisfactory**	**Exemplary**
Identification Determining necessary information Discovering what is already known	Prepared limited list of applicable literature to search	Prepared broad list of applicable literature the team plans to search for their literature review Prepared list of possible information sources to locate information	Prepared a comprehensive list of applicable literature the team plans to search for their literature review Prepared a complementary list of information resources they plan to use in locating relevant information Created a plan to centrally record information that they learn they will need to create for themselves in the experimental phase
Acquisition Evaluating information Storing information Describing information	Created a shared citation manager account	Created shared citation manager account Created a plan to record the relevancy of individual information resources Created a plan to record how and where information was located	Created a shared citation manager account Created a description of a defined evaluation system to note the relevancy of information resources Created a detailed plan to note how and when information was located providing all the information to include
Utilization Locating team information Applying information	No plan created for adding new information outside of the literature review	Created a plan to store information created throughout the design process	Created a detailed plan to store information created throughout the design process, including storage location, file naming convention, etc.

assistance to individual teams to revise and strengthen their plans.

For longer-term assessment to guarantee that the instruction impacts the students' behavior and work processes, the most effective assessment technique is to add the instructor and librarian to each design team's collaborative citation manager group. The instructor and librarian can then periodically check each group's progress and provide formative assessment throughout the entire design process. The instructor and engineering librarian can monitor rates of adoption of the techniques outlined as well as make just-in-time suggestions for improvement to each group's methodology. This approach also allows the engineering librarian to learn what information-seeking skills might need further development and provide additional instructional interventions at the point of need.

The viability of this method of assessment would depend on the size of the design classes and the overall workload of the engineering librarian. (An engineering librarian supporting multiple departments' design classes at a large research university would quickly find him- or herself overwhelmed.) Along with the volume of groups requiring observation, this method of assessment would require supervision over the project's entire life span.

A less time-intensive assessment process would be to check in with each group in a more informal manner, via e-mail or by dropping in on a design team meeting, to learn where they've searched, what they've found, and how they are storing and sharing their information and to discover any outstanding information needs they still possess. For both of these longer-term assessments, conducted throughout the project's life cycle, the information management plan produced by the student groups would serve as a gauge for assessing success.

Another, less direct, way to assess the impact of the instruction on student behavior would be to send out a survey at the end of the design project asking students to share how they managed their information. This assessment method, although less of a time burden, relies on student memory and does not provide an opportunity to intervene and augment student behavior as it unfolds.

EXPANDING THE SKILL SET

As previously mentioned, the best practices of information management laid out in the citation management exercise—having an agreed upon process for adding information, critically assessing the information gathered, and the importance of transparency and strong communication—can be expanded throughout the design process. Information management is integral in collecting data from experimental models, gathering stakeholder feedback, and reporting out findings to stakeholders.

Because the underlying skills are the same, the example featuring citation managers outlined earlier could be supplemented or repeated with a similar exercise using other collaborative resources. The central idea, using a tool that will eventually save students' time to capture their attention and ensure buy-in, remains the same. Similar to the example provided earlier, the instructor provides information on the basics of knowledge integration (and possibly project management documentation) and then has the teams apply it to their own beginning work plan. Instead of using citation management software, students could engage with a variety of software programs available to them for collaboration (Google Drive, OpenOffice, OpenProj, SharePoint, etc.). The same basic outline described previously for the citation managers would work here as well, with the instructor imparting the best practices of information management in examples and demonstrations of each tool before having the class experiment and report back on which features worked or were lacking in the different tools.

The same approach can also be applied to the creation of a data management plan to identify, acquire, and utilize the information created by the student groups. This reinforcement provides valuable scaffolding for the students, repeating important core concepts in information management practice. It also allows the instructor to go deeper into the importance of keeping good records of the information that the teams create, and how the management of those findings may prove important in other aspects of the design phase and ultimate manufacture.

Similar assessment strategies are appropriate when applying information management techniques to other portions of the design process. Using the information management plan created to conduct their literature review as a model provides students with a clearer understanding of the information management components of a data management plan and other future documentation.

SUMMARY

Information gathering and management occurs throughout the engineering design process, including searching the engineering literature, recording experimental data, and communicating with teammates and stakeholders, but it is vital for the design team to address this topic early in the design process to situate the team for maximal efficiency and ultimate success. Having students coordinate and collaborate on searches of the engineering literature for examples of prior art, current research in the area, and standards and regulations lends itself to the integration of information literacy skills into the information management process. Citation management software opens the door to an engineering design class's interest, with its promise of time savings and reduction in the duplication of work, to introduce information-literate management techniques. The successful use of these tools to employ information-literate information management practices illustrates a model of general information management techniques that will inform the students' understanding of other aspects of data gathering and management in the team's design process.

SELECTED EXERCISES

Exercise 6.1

Break students into their design teams and have them create a shared citation manager account. Instruct them to brainstorm places to look for literature on their design topic, find at least three citations, and practice importing them into their shared account. Once students have some citations loaded, have them devise a plan for organizing their citations within the citation manager's structures (i.e., determine what types of groups or folders they want to create to organize their citations). Also have the students discuss how they will evaluate and communicate about the citations they add using the tags or notes features. After students have conceived a plan, reconvene the larger group and have the different teams share their plan and allow their classmates to provide feedback.

Exercise 6.2

In their design groups, have the students come up with a shared space to save other pieces of information they plan to gather during their design project (e.g., Google Drive, Dropbox). Have students devise a folder structure and file naming conventions to make the retrieval of their created information intuitive and efficient. After students have devised a draft, have them share their organization plans with the larger class.

ACKNOWLEDGMENTS

The ideas represented in this chapter have evolved out of the development of a workshop that I created in collaboration with my colleague, Jody Kempf, Instruction Coordinator for the Science & Engineering Library at the University of Minnesota, Twin Cities, and the work I did adapting that workshop for in-class integration for a mechanical engineering design class, partnering with faculty member Dr. Susan Mantell.

REFERENCES

Bredillet, C. (2004). Projects: Learning at the edge of organization. In P. W. G. Morris & J. K. Pinto (Eds.), *The Wiley guide to managing projects*. Hoboken, NJ: John Wiley & Sons, Inc. (pp. 1112–1136).

Childress, D. (2011). Citation tools in academic libraries: Best practices for reference and instruction. *Reference & User Services Quarterly. 51*(2), 53–62. http://dx.doi.org/10.5860/rusq.51n2.143

Duong, K. (2010). Rolling out Zotero across campus as a part of a science librarian's outreach efforts. *Science & Technology Libraries*, *29*(4), 315–324. http://dx.doi.org/10.1080/0194262X.2010.523309

Kraaijenbrink, J., & Wijnhoven, F. (2006). External knowledge integration. In D. Schwartz (Ed.), *Encyclopedia of knowledge management*. Hershey, PA: IGI Global, pp. 180–187. http://dx.doi.org/10.4018/978-1-59140-573-3.ch024

Kuhlthau, C. C. (2004).*Seeking meaning: A process approach to library and information services* (2nd ed.). Westport, CT: Libraries Unlimited.

O'Sullivan, C. (2002). Is information literacy relevant in the real world? *Reference Services Review*, *30*(1), 7–14. http://dx.doi.org/10.1108/00907320210416492

Presidential Commission on the Space Shuttle Challenger Accident (1986). *Report of the Presidential Commission on the Space Shuttle Challenger Accident. Chapter V: The contributing cause of the accident*. Retrieved from http://history.nasa.gov/rogersrep/v1ch5.htm

Singh, J. (2008). Sense-making: Information literacy for lifelong learning and knowledge management. *DESIDOC Journal of Library & Information Technology*, *28*(2), 13–17.

Statt, D. A. (2004). *The Routledge dictionary of business management* (3rd ed.). London: Routledge.

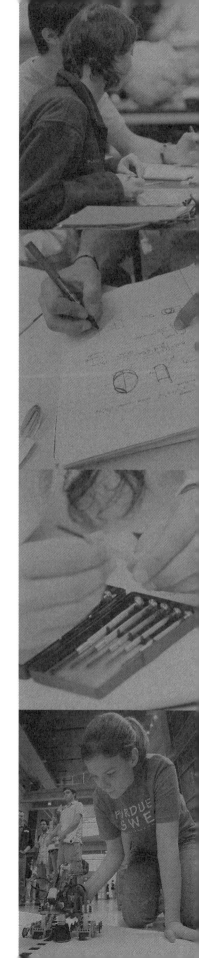

CHAPTER 7

FIND THE REAL NEED

Understanding the Task

Megan Sapp Nelson, Purdue University

Learning Objectives

So that you can guide student design teams to find the real needs of clients, upon reading this chapter you should be able to

- Distinguish between different types of stakeholders in a design project, in particular between client(s) and users
- Describe the common challenges that student design teams face in identifying and capturing the full range of needs, wants, and expectations of various stakeholders
- List and describe the benefits of a user-centered approach to developing project requirements and constraints
- Demonstrate how active information gathering techniques reveal the needs and wants of project client, users, and other stakeholders

Clarify the Task

INTRODUCTION

Once the team is organized and a code of conduct has been agreed upon, team members are ready to explore the design task. This usually commences with a design brief that contains the client's initial interpretation of the problem to be solved. However, a project team that considers only the design brief may substantially miss the mark in their design solutions. This is not only because only so much information can be communicated in a written document, but also because often clients do not know what exactly they want. This can be because they are unaware of possibilities or because they themselves have incomplete information about the needs of different stakeholders in the project.

Stakeholders are central to the design process. They are any individual who has a vested interest in the outcome of the project. That interest may be of a financial, utilitarian, or social origin. Stakeholders may provide funding for the process, specify problems that must be resolved or improved in the resulting solution, and influence both the scale and the time frame for a given project.

Stakeholders have both needs and wants that have to be captured, analyzed, and transformed into a set of requirements (those functions and features that must be present in the final artifact). They may also be a source of constraints, limitations placed upon a design project by any of a number of factors, including available resources, environment, legal requirements, and societal impacts. There are a few different kinds of stakeholders who are important to the design engineer (see Figure 7.1.) A client is a stakeholder who requests that an artifact be developed—that is, the entity that is paying the bills for the project. A user is a stakeholder who interacts with the artifact at any time during its life cycle, generally with the purpose of taking advantage of its features.

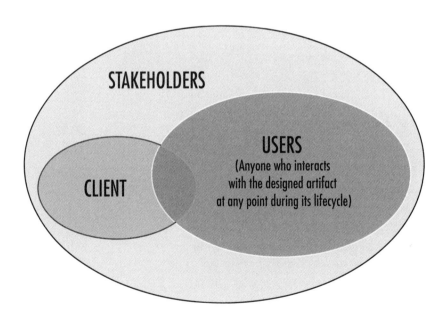

FIGURE 7.1 Stakeholders, clients, and users.

While clients make the investment of resources (time, money, personnel) to initiate a design project, they are not the only people impacted by the design process and the resulting artifact. Customers of the product, other end users, community members, maintainers of the artifact, and those who will ultimately dispose of the artifact when it has exceeded its natural life are all stakeholders in the design process.

The process of designing with the end user in mind is called *human-centered design*. The International Organization for Standardization's (2010) ISO 9241-210:2010 lists the following benefits for adopting a human-centered design approach:

a. Increasing the productivity of users and the operational efficiency of organizations;

b. Being easier to understand and use, thus reducing training and support costs;

c. Increasing usability for people with a wider range of capabilities and thus increasing accessibility;

d. Improving user experience;

e. Reducing discomfort and stress;

f. Providing a competitive advantage, for example, by improving brand image;

g. Contributing towards sustainability objectives. (p. 4)

Central to the human-centered design approach is the need to elicit information from stakeholders. Effectively eliciting information from others requires strategies and tools not often covered in the engineering curriculum, first to identify who might be a client or stakeholder in the project, and then to retrieve relevant information from those individuals. This chapter provides guidance for gathering useful information from a variety of stakeholders for the development of design requirements and constraints.

COMMON CHALLENGES FOR STUDENTS

Eliciting information from the design client and other stakeholders is a significant challenge even for experienced engineers. For students, it can be highly frustrating. The challenge for the engineering designer lies in drawing out the design client's understandings and observations and comparing that information to ideas elicited from others in order to get a comprehensive picture of the existing environment, the identified problem, and the most desirable outcome. Constructing this knowledge relies heavily on communication skills, not as taught in undergraduate speech classes, but as practiced on the library reference desk and other public service points. These interactions often require extensive interaction and follow up to tease out the client's fundamental question, let alone the final answer. Most undergraduate engineering design students will need to be explicitly taught skills to enable them to perform this type of interaction (Nelson, 2009).

> You can illustrate the challenges of communication to your students with an icebreaker used to build communication skills. Two individuals sit back to back. One individual is given a piece of paper with an abstract geometric drawing. The person holding the paper describes the abstract geometric figure to his or her partner. The partner then draws the figure as he or she believes that it has been described. The outcome frequently looks very little like the original drawing. In many ways, this icebreaker illustrates the challenges of accurately communicating design specifications and requirements.

Clarify the Task

Undergraduate engineering students are frequently accustomed to having all the relevant information presented to them, in the form of course textbooks, lecture notes, and supplementary materials. Such passive information acquisition does not work in the context of an open-ended design project. It is simply not possible for the design client to provide all necessary information to the design team in a single interaction, or even many interactions (Damodaran, 1996). The student designer needs to develop active information gathering skills, so that they have the ability to seek out important issues and relevant information that are not presented to them. Students frequently struggle with this change in their learning experience and consider it annoying, frustrating, and difficult (Zoltowski, 2010). Practicing active information gathering in prior course work can increase student abilities to adjust to the active information gathering that is necessary for design success.

Gathering user input can also be challenging for students because the information is not always direct or consistent, and the stakeholders may not be able to articulate their needs explicitly. They have latent (hidden or unknown) knowledge of the system or the problem that they might never have considered on a conscious level: "Oh, of course, we always put the peanut butter on before the jelly" (Vokey & Higham, 1999). And they may be able to identify that an aspect of the design project isn't in accordance with their understanding of the situation but are unable to articulate the specific ways that it does not mesh with their worldview: "It just doesn't feel right, I can't describe it." The engineering designer needs to understand the situation being described by the client and translate the client's observations into a design deliverable that interfaces well with the existing environment that the client

works in, as well as fixing or eliminating existing problems. Students need practice turning an initial statement, such as, "I need a pencil and paper," into a functional need, such as, "I have to communicate with others in a textual/graphic manner."

Students will also need to learn how to engender an open mode of communication to facilitate access to latent information. For the engineering designer, establishing a relationship with the client and providing prompt responses to suggestions or concerns raised helps create an environment in which the client feels comfortable sharing ideas, perspectives, and uncertainties. The initial client discussion should not be thought of as a one-time meeting but rather as the opening contact point in an ongoing relationship. If the design team does not maintain effective communication with the client and indeed other stakeholders after an initial meeting, it is much more likely that the artifact they design will not meet expectations or the real needs and consequently need extended revisions (Zoltowski, 2010).

Finally, it is critical to recognize that the client and the engineering designer may talk about the problem and possible solutions in quite different ways; the former in everyday language and the latter in technical terms that might not be understood by a lay audience. In other words, engineering as a discipline and an engineer as a practitioner must be aware of their use of words in particular and privileged ways. If a word is not clear to the client, the client may not ask for clarification to avoid looking unintelligent to the designer. In that way, important clarifications are missed and crucial opportunities to build mutual understanding between the client and designer are overlooked. Designers should target their language to the level of a senior in high school. This is slightly more sophisticated language than used in popular media, but much less sophisticated than used in an academic journal.

EXPLORING CLIENT BACKGROUNDS

Prior to meeting with a client, it is important to seek out basic information that will assist engineers in understanding the context of the client. That context may include motivations, available resources, goals, and financial information, as applicable. The request for consultation received from the client may be either vague or specific but generally does not give much context. The website, mission statement, strategic plans, and newsletters or press releases detailing recent developments within the organization are the first place to start gathering information about the client. These resources detail factual information, as well as provide insight into the organization's goals and culture. The resultant product will have to perform successfully within the setting and culture of the organization, so this information provides important context for the design project.

Another important corporate document is an organizational chart. This helps the designer understand what part of the organization the task is being solicited from, what other departments the project will likely impact, and potential additional stakeholders to interview. Having a basic understanding of the organizational structure will assist the designer in collecting and understanding information from the stakeholders.

Once company-specific documentation has been examined and understood, the next step is to look beyond the organization for additional information. Public information about the client organization can come from a variety of sources. Newspaper articles generally are tied to press releases and will contain information

similar to the internal documentation. However, they can be valuable for getting a community perspective of the project stakeholders. Newspaper articles can also uncover ethical contexts that the design deliverable will exist within. For example, if a newspaper article highlights how a client is dealing with privacy issues in the online environment and the design solicitation is for an online application, the team needs to clarify that aspect with the client.

Government documents provide insight as well, particularly when the client is a public corporation that must file quarterly and annual financial statements. These statements can give insight into emerging areas of growth for the organization, areas that are less competitive, and the available resources that the organization may draw on to support this project. For more on gathering information on the external context of a design project, see Chapter 8.

ELICITING INFORMATION FROM CLIENTS AND OTHER STAKEHOLDERS

In terms of the engineering design process, clients represent a significant source of specialized knowledge; they have unique knowledge and expertise related to the design context, as well as insights into the needs, wants, and constraints of the project. In the course of their day-to-day processes and activities, clients provide insight into what works well, what does not work, idiosyncrasies of any systems or technology currently used, and local cultural or organizational expectations. Clients and users are seldom consciously aware of some of the particularities of the work processes in their organization. They generally just go about their activities, carrying them out as they normally would without

Clarify the Task

extensive consideration regarding how and why a process works or does not work. Thus much of their knowledge is tacit—hidden and thus difficult to gain access to (Polanyi, 1966). It is knowledge similar to how to ride a bike or perform a similarly complex manual task.

Related to this is latent knowledge—that is, things generally known but not under conscious control of the individual (Vokey & Higham, 1999). Latent knowledge may be experienced as a gut feeling or just a part of everyday life that, when changes or violations emerge, the individual may say just doesn't feel right (Gorman, 1999). This cumulative wealth of tacit, unrecorded knowledge of clients and users includes information that will determine whether a design project is ultimately successful in the long term.

For designers, eliciting the tacit and latent knowledge of their clients is a significant challenge. Each individual client and stakeholder has a unique perspective that may influence the determination of design requirements and constraints. In particular, as experience, job responsibilities, and personality vary, so do the observations that individuals make and the resulting understanding that they have of how the project design will impact and interact with current practice. There are multiple methods for retrieving this information. Interviewing can be used to assist the clients to think in new ways about what they know. Observation can identify behaviors and patterns that the clients don't even realize exist.

IDENTIFYING STAKEHOLDERS FOR INFORMATION GATHERING

Success in design depends heavily on successfully eliciting the knowledge that stakeholders have accumulated through experience, obser-

vation, and other institutional knowledge that they maintain. But who are the stakeholders—beyond the client and people who will use something that is designed? Brainstorming a list of everyone who could potentially come in contact with the artifact to be designed is the first step to developing a comprehensive information collection plan. Personnel lists and organizational charts may provide insight into who should be asked for information. Identifying a specialist insider (e.g., a secretary, a manager, a supervisor) who sees the big picture of the organization as well as the work flow that occurs daily can be invaluable for determining who should be asked for input in the design process.

If possible, observing the clients, users, and other stakeholders in the operational environment in which the artifact will be used provides access to information that may not be available in any other way. In a demonstration of this technique for a news magazine story, the design firm IDEO went to a grocery store and observed shoppers. The firm determined that professional shoppers went about the process of shopping in a different way than household shoppers. The professional shoppers were much more efficient, and the key to their efficiency was to leave the cart at the end of the aisle so that there was no possibility of getting caught behind slowly moving household shoppers. This influenced the ultimate design of their cart (ABC News Nightline, 1999).

Observation is a time consuming but flexible model for identifying individuals who possess latent information and then collecting that information. Noyes and Garland (2006) provide a short overview of observational practices. Observations can be designed so that the observer is either covert (not engaging the subjects of observation) or overt (interacting and asking questions with the subjects). A good plan for an observation (Noyes & Garland, 2006) attempts to answer the following:

- Why?
- Who? (All or a selection of stakeholders?)
- What? (Define the behavior to be focused on.)
- Where? (Define the physical boundaries.)
- When? (Define the overall appropriate temporal parameters.)
- Duration? (Define the sampling method.)
- How? (Define the type of recording.)
- Role? (Define the researcher's level of participation.)

The primary advantage of observation is the immersive nature of the process. It helps the designer become familiar not only with the client and users in their work context but also with the environment, including stakeholders, organization-specific work flows, and the exceptions that are evident only in the environment where the design deliverable will be introduced. Immersion within the environment (even if only for a few hours) combined with in-depth interviews gives a deeper understanding of the situation and constraints for the design project than an interview alone.

INTERVIEW TECHNIQUES

A design project is generally initiated at the request of the client. Multiple meetings with the client help tailor the client's vision of the project into actionable information. An interview plan is an important tool to improve the efficacy and efficiency of a client meeting. Based on the questions typically asked of journalists—who, what, when, where, and how—a planned interview provides the interviewer an opportunity to brainstorm potential topics of discussion before

the meeting, organize the interview so that it flows well, phrase the requests for information in an open-ended manner so as to draw out the knowledge the client has, and create a document that structures notes taken and reminders for follow ups at a later time (Nelson, 2009). Figure 7.2 provides an interview plan that was developed for Engineering Projects in Community Service (EPICS) at Purdue University.

The planned interview not only focuses on open-ended questions but also encourages the interviewer to strategically design the interaction to foster the outcome of the interview. Active listening, a process that encourages critical consideration and follow up on statements at the time of the interview, is made easier by having a plan for the interview. It allows the conversation to be redirected back toward the goal the interviewer has in mind. Active listening requires vigilance during the interview. Including questions that will check the perceptions of the interviewee is important for developing a common understanding of the problem and eliciting more detail (Nelson, 2009). Perception checking is a process by which the engineering designer verifies his or her understanding of what the interviewee has said by rephrasing the question—for example: "If I understand you correctly, the file is then sent from you to someone in quality control for testing." This allows the interviewee to confirm, deny, or augment what was previously said. This type of language does not come naturally, so perception checking must be practiced in order to enable successful, smooth implementation during an interview.

It is very important to keep a detailed record of what transpires within a client interview. Video or audio recording provides the most complete record. However, indexing or transcribing the resulting file generally requires specialized software and trained transcribers.

CLIENT INTERVIEW PLAN

Team: _____ Project name: _____
Team member: _____

Client Description
Client: _____
Organization mission: _____
Primary stakeholders: _____

Interview Questions
(These are not in order that they will be used in an interview. These are just suggested questions to begin the interview process.)

How:
...do you envision using this product?
...are similar products currently used at the project partner organization?
...is the task this product will replace currently carried out?

What:
...current problems will be solved by the product?
...are the specific functions of the product?
...resources are already available for creating the product?
...solutions have already been tried?
...environmental stresses or forces do the product need to withstand?
...safety guidelines must be taken into consideration?
...do you imagine could _____?
...have you thought of?
...would it be like if _____?

Where:
...have you seen a similar product to what you are envisioning?
...will this be located?
...do you envision housing this project?

Who:
...will be using this product?
...is most affected by the task that this product will contribute to?
...needs a (module, password, access)?

When:
...is this product most needed?
...is this product needed by?
...is this product most likely to be used?

Hints for a successful interview:
Attitude: Open attitude leads to open communication.
Attention: Show attention by body language.
Focus: Focus on content and ideas. Make mental notes of questions to ask when the speaker has finished.
Probe: Ask questions that will provide opportunity for more details to emerge.

FIGURE 7.2 Client interview plan.

Generally, permission of the interviewee should be requested prior to recording an interview, even if it is just a simple permission form presented to the client. Prior communication will avoid surprises so that the team does not arrive at the site only to be told that the company has a policy against recording.

In the case that audio or video recording capability is not available, or a permanent record is prohibited by confidentiality agreements (see Chapter 5), team roles should be assigned to ensure duplicate notes are taken and full coverage of the interview is captured. Multiple note takers should record not only the oral content of the interview, but also make notes of topics that body language and other cues indicate should be followed up on at a later time. For example, if a supervisor is the primary client and makes a statement, but a subordinate opens his or her mouth to speak and then closes it again, a note should be made to talk to that individual again at a later time about that specific topic.

As an interviewer, the engineering designer also must consider his or her own role in the interview. Body language on the part of the interviewer can send a message to the interviewee either that the interviewer is engaged in what the interviewee is saying or is bored and would rather be someplace else. Similarly, nervous habits such as clicking pens or tapping feet can give the impression of impatience or distraction. Practicing interviews ahead of time will help to make interviewers aware of their tendency toward these distracting actions. It is useful to have others on the design team brainstorm alternative ways that interviewers can deal with nervous habits.

If an interview is being conducted one on one, and the interviewee is having difficulty explaining his or her latent knowledge (the "it doesn't feel right" phenomenon), several different approaches aligned with the preferences of different learning styles may help to draw out the information that the client has in mind. Table 7.1 provides examples of strategies that might assist interviewers in eliciting information from informants according to their preferred learning styles. It uses the four dimensions of learning style based on Felder and Silverman (1988): active-reflective, sensor-intuitive, visual-verbal, and sequential-global. For example, walking a client who is an active, sensor, visual, and global learner through a physical space or work flow may help the client preferentially to see how a proposed solution might impact the current work flow.

At various time all people prefer to receive and deliver information in different ways. As Felder and Soloman (n.d.) observe: *everybody is active sometimes and reflective sometimes and everybody is sensing sometimes and intuitive sometimes.* It depends upon the circumstances, so it is critical not to pigeonhole informants into a set of characteristics. The designer should keep all the strategies at hand and deploy them as most appropriate, treating each informant as an individual with unique learning and informing styles.

Using Post-it notes to capture ideas from a group and then categorizing them by collating them on the wall or table may be helpful. Similarly, encouraging a client group to model or act out a work flow or process may provide additional insights as well. The client interviewee group can be split by similarities (IT personnel, sales people, etc.) and those groups asked to brainstorm the implications of the design solution for their department. Then, the client interviewees can be grouped across function (e.g., one IT person, one sales person, and one manager) and asked to brainstorm how the design task facilitates or hinders cross departmental communication and work flows. Using activities, drawing on visual and oral cues, and

Clarify the Task

TABLE 7.1 *Information-Eliciting Strategies Based on Informant Learning Style (Using Felder-Silverman Learning Styles Inventory)*

Learning Style	Key Characteristics	Eliciting Information Strategies
Active	Prefer doing something active; discussing or applying it or explaining it to others	Ask them to show you what they do. Invite them to talk you through it and to demonstrate in the authentic location
Reflective	Prefer to think about things quietly by themselves	After talking with them, offer them an opportunity to think about things (say, overnight) and suggest they write down their thoughts and send these to you later
Sensing	Prefer facts, details, practical matters, the "real" world	Encourage them to give you the facts as they see them; ask them to explain what is done and why
Intuitive	Prefer discovering possibilities and relationships	Ask them for their ideas about how things work around here Elicit their theory of what is happening and why
Visual	Relate best to visual information—pictures, diagrams, flow charts, time lines, films, and demonstrations	Get them to discuss what happens here using available operational charts, performance graphs, and the like
Verbal	Get more out of words—written and spoken explanations	Invite them to tell you stories about how things work here; these can be war stories of practice or anecdotes about the organization or the personalities therein
Sequential	Prefer linear steps, with each step following logically from the previous one	Ask them to walk you through what happens step by step and explain the rationale of why it is so or what has been tried previously
Global	Take large jumps; think almost randomly without seeing connections, but then suddenly get it	Encourage them to paint the big picture about the place Ask if they have a metaphor that captures what happens around here

Modified from Felder & Silverman, 1988.

group discussions will help the client or client team to fully consider what each person knows and to articulate their opinion(s).

Additionally, wire framing or concept mapping may assist the client or client team in categorizing and identifying their work flow. Talking through either of the previously mentioned approaches will assist them in articulating ideas about their work and processes.

After the interview, it is very important that the designer immediately return to his or her notes and/or recordings of the interview to confirm that the contents are unambiguous and that no major points were missed, and to add in any additional impressions or ideas that occurred to the engineer during the interview session. This can be as simple as a brief review of the notes, or as complex as a weighted decision matrix (see Chapter 11). If the interview was recorded and transcribed, the designer can annotate the print transcription where further follow up is needed. If a full transcription is not possible, the interview can be indexed by listening to it again, making note of the time stamp when a topic emerged, and noting the topic, as well as any additional follow-up questions.

Regardless of technique, the goal is to immediately return to the interview and add any emerging observations or questions into the written record for the project. A significant amount of value from the interview is lost as initial impressions and questions are forgotten over time. For future design team members, an accurate, extensive record created at the time of the interview is a valuable asset for the rest of the design cycle. A strong knowledge management system for the team will ensure that the information gathered remains accessible throughout the project, to maintain alignment with the determined needs.

PERSONAS

A useful exercise at the end of a group of interviews is the creation of personas. In this case a persona does not represent one person, but an archetypical user of the design deliverable. This persona helps draw together the major commonalities across multiple interviews and highlights specifications that will serve the greatest number of users. The personas then become living documents by which to test assumptions made by engineering designers and recall the human-centered part of human-centered design (Pruitt & Adlin, 2006).

In general, a persona looks a little like an online profile of a person. It includes a representative photo and sample characteristics, such as age, work roles, home life, immediate and long-term goals, and a description of how that archetype interacts with the design deliverable. For extended information on the process of creating a persona, see Pruitt and Adlin (2006). Creating personas is a quick way to summarize the pertinent information found during the interviews. Either way, the persona serves to recall the designers back to the specifications elicited from the interviews throughout the design life cycle. For further discussion of the use of personas, see Chapter 8.

ADDITIONAL TECHNIQUES

IDEO has created a deck of cards (http://www.ideo.com/work/method-cards) that contains 50 strategies for eliciting information based on four approaches—learn (from what already exists), look (at what people do), ask (people), and try (out an idea). Comparable strategies are published by the d.school at Stanford (http://dschool.stanford.edu/use-our-methods). These and similar toolboxes of need-finding and knowledge-eliciting techniques can be used as a resource for a design class to not only prompt students to learn and adopt creative new approaches to get a more comprehensive information background on their project, but also teach the students to become creative design thinkers.

SUMMARY

In this chapter we considered the information that our stakeholders possess regarding our design project. We looked at several techniques that allow us to access that information and gather it for the creation of design require-ments and constraints. Using the information gathered by users who are clients and stake-holders in combination with the information gathered from external sources (see Chapter 8) allows the engineer to understand the problem more deeply, refine the requirements, and iden-tify constraints. These are then used to create the design specifications that will guide the cre-ation of solutions to the design problem.

SELECTED EXERCISES

Exercise 7.1

Have students brainstorm five to six potential sources of information about the organization they are working with that *were not* authored by someone in that organization. Have them search these sources for information. Ask them to discuss what they found and how the infor-mation produced by someone outside the orga-nization differed from the corporate authored materials. Have them evaluate the strengths and weaknesses of both types of information.

Exercise 7.2

Practice perception checking using the follow-ing exercise.

Have one individual in a student team speak for two to three minutes on a topic with which they are familiar. Examples include changing a bicycle tire, baking a special dessert, playing an instrument, building a website, programming in a specific language, gardening, and so forth. Have the other members of the team listen and write down follow-up questions phrased to check perception. For instance, a student might ask a speaker on the topic of changing a bike tire: "If I understand correctly, you are

matching something about the tube to the tire. How do you know which tube goes with that tire?"

Exercise 7.3

Students learn to recognize their own body language and verbal ticks when they are made aware of them either by videotaping or by hav-ing peers provide feedback. Videotaping a mock interview, with students taking on the role of both interviewer and interviewee, allows the students to objectively understand how their communication skills appear to others. (This can even be done with a simple smartphone.) This is best done in a small group rather than as an entire class. If possible, the students should take turns interviewing and being interviewed so that every person plays both roles. Those who are acting as interviewer should plan the interview with the goal of eliciting specific in-formation. Provide feedback on body language and word choice and expose students to alter-native interview techniques they may use to get similar or better quality information.

ACKNOWLEDGMENTS

Special thanks to David Radcliffe for creating the mapping of information-eliciting strategies to the Felder and Silverman (1988) preferred learning style of the informants (see Table 7.1).

REFERENCES

ABC News Nightline. (1999). *The deep dive*. ABC. Retrieved from http://www.youtube.com/watch?v=JkHOxyafGpE

Damodaran, L. (1996). User involvement in the sys-tems design process-a practical guide for users. *Be-*

haviour & Information Technology, 15(6), 363–377. http://dx.doi.org/10.1080/014492996120049

Felder, R. M., & Silverman, L. K. (1988). Learning and teaching styles in engineering education. *Engineering Education, 78*(7), 674–681. Retrieved from http://www4.ncsu.edu/unity/lockers/users/f/felder/public/Papers/LS-1988.pdf

Felder, R. M., & Solomon, B. A. (n.d.). *Learning styles and strategies*. http://www4.ncsu.edu/unity/lockers/users/f/felder/public/ILSdir/styles.htm

Gorman, M. E. (1999). Implicit knowledge in engineering judgment and scientific reasoning. *Behavioral and Brain Sciences, 22*(05), 767–768. http://dx.doi.org/10.1017/S0140525X9936218X

International Organization for Standardization. (2010). *ISO 9241-210:2010, Ergonomics of human-system interaction—Part 210: Human-centred design for interactive systems*. Geneva, Switzerland: Author.

Nelson, M. S. (2009). Teaching interview skills to undergraduate engineers: An emerging area of library instruction. *Issues in Science & Technology Librarianship, 58*. http://dx.doi.org/10.5062/F4ZK5DMK

Noyes, J., & Garland, K. (2006). Observation. In W. Karwowski (Ed.), *International encyclopedia of ergonomics and human factors* (2nd ed.). 3 volume set. Boca Raton, FL: CRC Press. http://dx.doi.org/10.1201/9780849375477.ch635

Polanyi, M. (1966). *The tacit dimension*. Garden City, NY: Doubleday.

Pruitt, J., & Adlin, T. (2006). T*he persona lifecycle: Keeping people in mind throughout product design*. Boston, MA: Morgan Kaufmann.

Vokey, J. R., & Higham, P. A. (1999). Implicit knowledge as automatic, latent knowledge. *Behavioral and Brain Sciences, 22*(05), 787–788. http://dx.doi.org/10.1017/S0140525X99582186

Zoltowski, C. B. (2010). Students' ways of experiencing human-centered design. (Doctoral dissertation). Retrieved from ETD Collection for Purdue University.

Clarify the Task

CHAPTER **8**

SCOUT THE LAY OF THE LAND

Understanding the Broader Context of a Design Project

Amy Van Epps, Purdue University
Monica Cardella, Purdue University

Learning Objectives

So that you can guide student design teams on real needs of clients, upon reading this chapter you should be able to

- Identify a broad range of factors to consider in understanding the context of the design solution, including geographical, economic, and cultural factors and human, material, and environmental resources

- Identify processes and sources for learning more about the context of the design task

- Synthesize the information that is collected into a form that is useful

- Use information about the context to develop clear and measurable criteria for the design task

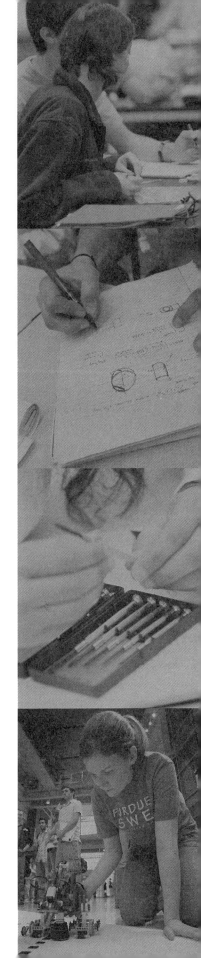

INTRODUCTION

In the previous chapter the importance of gathering information from stakeholders was discussed. However, in order to clarify the task more fully, designers need to also take into account the contextual components of the artifact being designed, such as the geography, economy, cultural norms, material resources, human resources, and environmental resources. This information helps the team create a coherent and cogent description of purpose and a scope of the design need or opportunity for a particular problem in a specific location. After collecting this information, the design team establishes a set of criteria by which possible alternative solutions are evaluated and compared (Chapter 11), and the final outcome is judged by the client, user, and other stakeholders (Chapter 13).

This chapter will focus on working with students as beginning designers who are attempting to develop informed design practices, by guiding the students to explore, comprehend, and frame the problem thoroughly. Building on the techniques of gathering client information presented in Chapter 7, the exploration continues into areas where the users or stakeholders may or may not have information to share. These issues may not come to mind for the users during interviews either because they are so immersed in the environment on a daily basis that they do not see the details and possible design problems, or because they are located in a different area and are unaware of issues related to a particular location.

COMMON CHALLENGES FOR STUDENTS

Beginning students often take a narrow view of a design project, considering it a technical task rather than a human undertaking with social

and environmental consequences and considerations. A common description of an engineer is, indeed, a problem solver. However, this is a limited vision of an engineer. Too often students focus on the solving part of design work, rather than deeply understanding the problem. As a result, they might end up solving the wrong problem, or develop solutions with critical errors because a particular constraint was not well understood. It may be that they don't recognize the importance of understanding the broader context, or that they don't have the necessary tools to do so. We do know that female engineering students seem to be more concerned about the broader context than their male counterparts as freshmen, but this gender difference disappears by the time they finish college (Kilgore, Atman, Yasuhara, Barker, & Morozov, 2007).

As an example to illustrate these challenges, imagine that you were asked to design a playground for your neighborhood. What are all of the different things you would consider? What types of information would you want to have? Now imagine that you were asked to design a retaining wall system to prevent flooding of a large river. What are all of the different factors you would consider in this case?

Kilgore et al. (2007) found that students tend to think about a relatively short and narrowly focused list of things they would consider in designing a playground, types of information needed for designing a playground, and factors for designing a retaining wall. For example, for the playground problem, students mostly considered the overall cost of the playground, the safety of different activities, and the amount of time it would take to create different pieces of equipment. In a related study, Atman and her colleagues found not only that students who made more information requests and gathered more types (categories) of information tended to have higher-quality solutions (Atman, Chimka, Bursic, & Nachtmann, 1999), but that the number and variety of information requests increased with experience as measured in populations of first-year students, seniors, and professional engineers (Atman et al., 2007). In contrast to the three main types of information requested by novices, advanced students and experts considered information related to all of the following: accessibility, safety, material costs, budget, material specification, information about the area, labor availability and costs, body dimensions, utilities, technical references, legal liability, maintenance concerns, neighborhood opinions, neighborhood demographics, availability of materials, and supervision concerns.

In another study, Wertz, Fosmire, Purzer, and Cardella (in press) analyzed reports students created for a design project for a first-year engineering course to investigate the types of sources students access while working on design projects, the students' ability to cite the sources appropriately, and students' ability to use information appropriately (i.e., to use information that is relevant and to use information to support their reasoning). The results from this study show that students mostly relied on Web resources and that their documentation skills were weak. However, when students did successfully document information, it was generally used appropriately. Thus, two other challenges for educators are (1) to prompt students to make use of many different types of resources, not only electronic ones; and (2) to reinforce documentation skills (such as using APA, MLA, or CBE format). This might be a matter of reminding students that these skills are not only relevant for their English or communication classes but also are important in their acculturation as ethical, professional engineers (see Chapter 5).

WHAT INFORMATION IS IMPORTANT? WHY?

Professionals (such as engineers, lawyers, doctors, and nurses) look for information based on specific needs (Leckie, Pettigrew, & Sylvain, 1996), and research shows that professionals consider many more needs related to a project than do novices. It is critical for novice engineering designers to understand and recognize which facets of the problem require additional information before they jump into generating solutions (Bursic & Atman, 1997; Crismond & Adams, 2012). Finding the right sources of information helps fill the knowledge gaps in any design project. It is also important for designers to realize that information gathering is a process that is likely to be revisited throughout a project as the team explores possible solutions and continues to interact with the clients and other stakeholders. Categories of information that influence design include geographical, economic, and local and cultural contexts of the problem. Design teams should also look at availability of resources, both human and material, in the location where any potential solution will be implemented.

REALITY CHECK 8.1

A team of engineering students was given a project to provide a play space in Ghana. They started to brainstorm solutions, figuring out what they could build out of mud, twigs, grass, and animal skins. They were quite surprised when introduced to the community to find it had modern tools and even (intermittent) electricity.

They students hadn't bothered to figure out what materials were available, if the project had a budget, or the types of play activities that were common in Ghana. What should the students have done differently?

Clarify the Task

Revisiting the playground example, there are many types of resources that will help the student get a complete understanding of the problem and the context for the solution. Some examples of contextual information include city or county building and zoning ordinances, culture of the community near the proposed location, budget, existing site conditions (grass, asphalt, pitch, drainage), local climate, and accessibility of the site for workers and future users. Various questions or considerations around budget can produce additional constraints or opportunities in a design project, be it finding additional or different equipment, or using a contractor or local volunteers for construction and/or installation.

For the retaining wall example, historical information that could be helpful in making design decisions includes water levels and volume of the river in question, history of flood and high water mark, frequency of flooding, seasonal variations in water flow, type of land, and occupants of the floodplain (e.g., farmland, petroleum refining plant, other manufac-

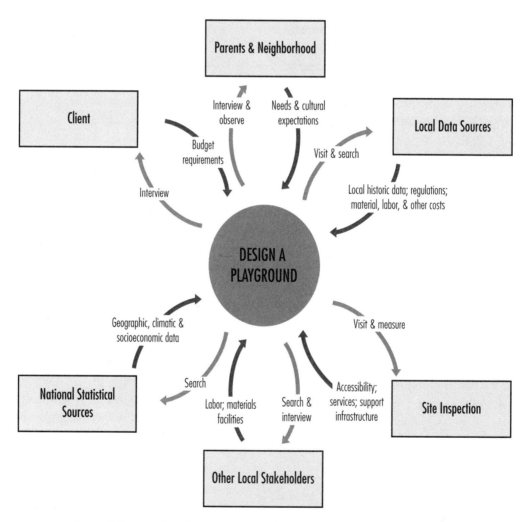

FIGURE 8.1   Relational diagram for information needs of the playground design project.

Clarify the Task

turing that could cause water contamination). Additional potential concerns include type of soil along the river and how easily it erodes, any communities or landowners who would be affected by the retaining wall, any aesthetic issues or concerns about the wall, and landowners whose property may be impacted. Human resources include the level of education/training of people involved in the project. (The information to be gathered from and about the people/clients related to the problem is discussed in Chapter 7.)

One way to get a more sophisticated sense of the types of information that are necessary for a complete contextual understanding is to use a concept diagram. These diagrams look a bit like part of a data flow chart, helping map where information comes from and what sorts of information are needed in consideration of the design project. Figure 8.1 shows a context diagram for the playground example.

CONTEXTUAL INFORMATION

As discussed in Chapter 7, the client can explain why the design project is being carried out and potential users have the most direct understanding of the need and community expectations. It is likely that conversations with the client may generate context concerns unknown to the user. Designers need to make notes about these issues and make sure they gather as much context information as possible on those topics.

Every design project takes place in a specific *cultural* context. This includes the prevailing local socioeconomic conditions, which can be discovered by reference to national, regional, or local statistical data and studies. Aspects of the broader cultural conditions are implicit in the problem statement provided by the client, but

this needs to be made explicit. It is important to determine what practices are considered normal or are forbidden by local custom of the primary user population. In the playground design, is the local neighborhood culture one where the children regularly gather and play together with only a few parents watching the group, or is the practice more about a small number of children gathering with all parents being present? The culture of an area becomes very important when the designer is working outside of a familiar situation or when the site is remote and cannot easily be observed. When this is the case, information sources include published information about a given culture and input from people who have been to the location. It also includes information that may come up in the cultural review, such as whether the community has a pattern of recycling that needs to be supported or restrictions on the number of people that can occupy an indoor space based on the limits of the current air handling system.

Historical information is a resource for possible solutions that have been proposed by others for a similar situation. Finding what has been done before and evaluations of what did or did not work are all important pieces of information to have before moving on to making a design decision. Techniques and locations for gathering this kind of supporting information are articulated in Chapter 10.

Environmental considerations include geographical and climatic information. It is imperative for designers to fully understand the location, so one should not depend solely on the stakeholders but observe the location while the people are using existing facilities. If something appears different than what the users stated, one should go back and ask for clarification and gather external information about the area. The geographical context includes the physical conditions of the site and the nearby areas. For example, Is it an indoor or outdoor setting? Does the construction need to work with an existing structure or is it a new construction? Is the construction site easily accessible for people and any required machinery? Outdoor issues can include annual snowfall, rainfall, wind, or sunshine. These are all important considerations, particularly for outdoor constructions such as the playground. Part of understanding the location and developing design constraints includes determining any local building regulations and codes. For the playground example this could include setback from the road, materials and paints considered safe around children, or height restrictions.

Of course a core consideration is the *economics* of the project across its entire life cycle. Budgets for design projects need to contain much more than just the cost of materials for whatever solution is finally selected (see Chapter 12 for more on material selection). For the playground example, the designers need to know if the land is already available or whether a site still needs to be identified and land purchased. Beyond purchase of equipment or materials, there are construction and/or installation costs and landscaping to ensure proper drainage of the land, safety of the children, and aesthetics. Another cost frequently overlooked is a consideration of any ongoing maintenance fees for equipment, power fees for lights, or city water fees for restrooms.

Legal information includes any applicable building codes—state, national, or international—that need to be followed, along with any local ordinances. Local governments may have laws concerning road setbacks, building height restrictions, or zoning requirements about the type of use a particular space can support. Additional legal requirements may arise from the contract that was signed.

One additional context component that needs to be considered is *infrastructure*. This includes a variety of information that will provide both criteria for any design solution and opportunities or ideas unique to a particular location. Criteria will grow out of information about local utilities, availability of services, and costs to connect with an existing infrastructure as well as maintain an ongoing service. Opportunities are likely to arise from discovering local businesses and services that make the design solution easier to implement through locally sourced materials or more appealing to the community through safe walking access and nearby amenities.

Material data sheets and vendors of commercially available materials components are primary sources of materials cost, as outlined in Chapter 12. Additionally, local availability of materials may be a consideration, especially with the growing interest in sustainability. Using locally sourced materials or native species (in landscaping) can decrease the environmental impact of the artifact being designed. Local labor costs can vary by location and the range of specialized skills required. In a case like the playground, consideration can also be given to local volunteer labor that may be available for construction. The cost of transport to site and specialist equipment needed for construction (e.g., earth moving equipment or cranes) should be considered.

Locating Contextual Information

The design team will need to determine which of the categories discussed in the previous section—cultural, historical, environmental, economic, legal, infrastructural—are most relevant to their particular project as they develop a strategy for acquiring needed contextual information. Table 8.1 summarizes contextual aspects and types of information rather than specific items or sources. Later chapters in this handbook provide details about different sources and what kind of information they contain.

It takes time to find relevant and trustworthy information. Just like the design process, gathering context is not linear. Any of these contextual information gathering steps could uncover information that causes the designer to review a previous set of information and add detail. The more information that can be gathered, and the more understanding the designer has of the overall problem, the more complete and satisfactory the final designed artifact will be.

Assessment of Information Gathering/Context Setting

One method of assessing the quality of information gathering is through peer evaluation of mini-presentations of the design setting and concerns. In a design class, teams working on other projects can provide external perspectives and help identify gaps in the contextual setting. Students can also create a problem statement document, referenced appropriately, that reflects their understanding of the contextual considerations. This document can be used formatively as the first step in an iterative process of problem refinement.

USING CONTEXT IN FRAMING THE PROBLEM

Once a student (or designer or engineer) has gathered information about the larger context, that information needs to be used to inform design decisions. Two tools that can help in the process of synthesizing the information are scenarios and storyboards. A third related tool is a persona. Designers create personas to synthesize the types of information collected about users and stakeholders into a fictional person (where the key to the practice is that the persona is not purely fictional, because the creation of this "person" is based in the evidence of the collected data about the stakeholders). Chapter 7 provides an overview of this design tool; in this section we describe how personas are used with scenarios and storyboards.

Scenarios

To complement the personas that the designer has created to embody the information collected about the stakeholders, the designer can create a scenario to synthesize information collected about the larger context of the design project. A scenario can be understood as a short story, where the persona is the starring character, and the crux of the storyline focuses on the persona's interaction with the product or process being designed. However, it is essential that the short story is not based in pure fiction, but instead that the details come from contextual information. At times the designer might focus the scenario on the user's life or experience prior to the introduction of the new artifact that has been designed (and so the story brings to light the user's unmet needs), while at other times the designer might instead create the scenario of how the new artifact is experienced by the user. It is also common for the designer to create both types of scenarios, as a before-and-after set (Preece, Rogers, & Sharp, 2002; Rosson, & Carroll, 2001; Stone, Jarrett, Woodroffe, & Minocha, 2005).

A scenario can summarize and remind designers of the different factors they should take into account in their design process. Students can review the example scenario provided in Box 8.1 and list all of the factors they would take into account if they were designing a

Clarify the Task

TABLE 8.1 *Contextual Considerations and Information Sources*

Type	Example Design Information	Example Sources
Cultural (including socioeconomic)	Demographic data Average income; income distribution Local employment statistics Ethnic neighborhoods—cultural norms Residential vs. commercial spaces ratio Attitudes to public facilities	National Census Data Reports of state or regional agencies Bureau of Labor Statistics User community observation Observation (photographs, frequency counts)
Historical	Trends in use of public facilities Success of past public facilities	Local histories including oral histories Newspaper articles Residents of longstanding
Environmental (geographical; climatic)	Annual weather patterns; snowfall, rain, sunshine, wind Soil types	National Oceanic and Atmospheric Association U.S. Geological Survey
Economic	Ongoing maintenance costs; electric, water, repair Nature, properties and availability of local (indigenous) materials Availability of general and specialized skills Availability of other people to assist (e.g., volunteer labor)	Local energy company rate sheet Better Business Bureau listing of local contractors or specialists
Legal	Safety requirements Required setbacks from a road Contracts	Local and state building codes Local authority rules and regulations Contracts/agreements with clients
Infrastructural	Community waste options (recycling, composting) Local services—accessibility (walking, parking, construction equipment)	Local utility companies (water, electric, sewage, gas) Directory of local business and services

playground for this neighborhood. Their lists might include the following:

- The appeal of the playground. Will children want to go there?
- Location within the neighborhood. Will families walk or drive? How much parking is available?
- Places for parents to sit.
- Shade.
- Ability to accommodate activities for children of different ages, activities that children of different ages can do together, and activities that keep 10 to 15 children occupied at the same time.
- Bathrooms, and possibly water fountains.

Storyboards

An alternative way to tell the story is through storyboards. Storyboards are a series of images and captions that provide a more visual summary of key features of the context in which the artifact being designed will be used, and can also portray a step-by-step flow of events associated with the use of the designed artifact (i.e., what happens first, what happens next, what happens last). The images used in the storyboard could be photographs, sketches, or other created pictures (Rosson & Carroll, 2001; Stone, Jarrett, Woodroffe, & Minocha, 2005).

USING INFORMATION TO DEVELOP CRITERIA AND CONSTRAINTS

Ultimately, designers must determine the scope of the work to be done in order to address the initial problem brief. Creating sce-narios or storyboards can help them synthesize contextual information to make decisions about what is within the scope of the project and what is outside the scope. However, these are just two tools that can help designers to make these decisions.

As information about the larger context is analyzed and synthesized, and perhaps depicted through the use of scenarios and storyboards, the information ultimately must lead to the identification and creation of appropriate requirements and constraints. The *criteria* (which include the things that designers would like the artifact being designed to do, or to not do or be) are used to differentiate amongst different options, while the *constraints* (or *requirements*) are criteria that *must* be met for the artifact to

---

BOX 8.1

Example Scenario—Summer Break

It was six weeks into summer vacation, and Janelle was bored with her toys at home. *Mom, can we go to Chuck E. Cheese? I'm bored.* It was 10 o'clock in the morning, and the sun was shining outside. *It's such a nice day. Why don't we go to the park instead?*

During the spring, the neighborhood playground had been transformed into a pirate ship, with a climbing net taking children from the ground to the ship's floor, a telescope and steering wheel installed at the top of a lookout platform, and slides exiting the ship to the lifeboats. Janelle enjoyed pretending that she was a princess captured by pirates, waiting for a rescue party to come. Soon, Janelle, her mother, Nora, and her younger sister, Sasha, were on their way to the playground. Only five blocks from their house, the playground was an easy walk away (even if a bit slow, with three-year-old Sasha as part of the walking party).

Once they reached the park, Sasha's pace increased considerably as she attempted to keep up with her seven-year-old sister, who was eagerly climbing the net up the ship's side. Sasha's mobility and agility hadn't quite developed to the extent that Nora was comfortable with her climbing up the net like her sister, so Nora directed Sasha to the ramp on the other side of the ship that would allow Sasha to board safely. Nora sat down on one the benches facing the pirate ship and began to read the magazine she had brought along. Soon she began to wish she had brought along sunglasses and a hat as she was squinting while the sun continued to rise. *Grow trees, grow. A little shade would be nice.*

One of the articles in the magazine got Nora to thinking about Janelle's birthday next month—perhaps they could hold her party at this park. They could incorporate the pirate theme throughout the party. *Are there enough activities to keep at least 10 kids busy? There aren't any picnic tables; we could eat and have cake back at the house either before or after we play at the playground . . . or bring along blankets for a picnic on the grassy area.*

Nora was interrupted by Janelle. *Mom, Sasha needs to use the bathroom.* Unfortunately, that meant a trip home—and it would have to be a fast trip home to avoid a potty training accident. Janelle was going to be disappointed. *If only there was enough space to install bathrooms at this park.*

be a viable option. Good criteria have three main features: they are clear, they are measurable, and they distinguish the feasibility, desirability, and viability of options. For example, one might say that the artifact being designed must be culturally appropriate. This would be an example of a constraint that is not clear or measurable. It can be improved through the gathering and use of information related to the cultural norms of the design context. A constraint that is clearer and more measurable would be that the retaining wall should not displace any historical landmarks. Criteria guide initial idea generation as well as later decisions (as the designer chooses amongst possible alternatives). Chapter 11 describes methods of evaluating design alternatives against criteria. Table 8.2 provides examples of criteria derived from contextual information.

USING INFORMATION TO BEGIN IDEATION

The how-why diagram is a powerful tool for exploring the context of a given design task and for exploring a much wider solution space. Thus it opens up new areas and avenues for information seeking. Figure 8.2 is a how-why diagram that was constructed around the initial design question: What types of head impact protection can we design for students in class? It seems many were falling asleep and being injured as their heads hit the desk.

If designers simply tackle the design task as posed, then they are seeking ideas about *how* this problem might be solved. In this case, three possible solutions are suggested: (1) the Wake-Me,

TABLE 8.2 *Sample Requirements Derived From Contextual Criteria*		
Type	**Sample Criteria for a Playground Design**	**Constraints/Requirements**
Cultural	Amount of space for the most popular sport or social activity for that region	Include at least 60 × 100 yards of space for a soccer field
Historical	Improves upon existing playgrounds in the area, measured by the number of features included that were absent in unsuccessful playground designs	Includes at least one new feature
Environment	Amount of shade present to protect children from sun Number of existing plants and trees displaced (should be minimized)	At least 50% of the playground is covered by shade trees Meets federal environmental impact regulations
Economic	Cost of construction (should be minimized)	Maximum construction budget is $10,000
Legal	Amount of shock the surface under the playground could absorb	Includes a minimum of 6 feet of fall zones in all directions for play equipment over 20 inches high
Infrastructure	Quality and quantity of amenities (e.g., water fountain, bathroom, parking)	Includes access to drinking water

a device that senses the onset of sleep and provides a mild electric shock to wake students before their head hits the desk; (2) the Snooze-o-Matic, a type of airbag in students' notebooks that inflates upon impact; and (3) the simple solution that the students all wear crash helmets to class. Each of these concepts would require accessing a variety of design information. In turn each of these three solution concepts can be fleshed out to find out how they may be realized in practice. So for example, the Snooze-o-Matic might be made up of four subsystems: a frame, a power source, an airbag, and a trigger. In turn we could ask how might each of these subsystems be achieved, and so

on down to each component. Thus, asking *how* narrows the design thinking to move toward more and more specifics.

However, if instead of asking *how*, the designer asks *why*, then the nature of the design task opens up and so does the potential solution space and also the range of information that might be sought. In the example, if the designer asks why we are trying to provide head protection, he or she might see the more fundamental problem of avoiding injuries due to boring classes. Asking *how* this might be achieved opens up a number of possibilities, including eliminating lectures or making classes more engaging (i.e., tackles the source of

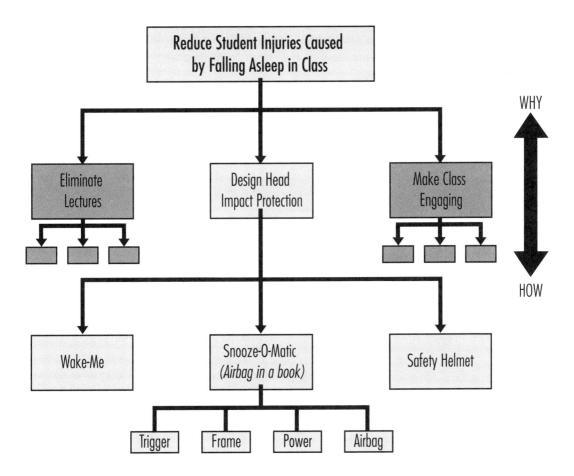

FIGURE 8.2   How-why diagram for head impact protection to prevent injuries when students fall asleep in class.

Clarify the Task

problem). Asking *how* either of these might be achieved poses a whole different set of design concepts, as indicated respectively by the solid darker blue boxes in Figure 8.2.

SUMMARY

A properly developed problem statement is just as valuable as the final solution. When presenting a solution, designers need to show not only what they are proposing, but why the solution meets the needs of the stakeholders and how the solution fits within the stakeholders' larger geographical, economic, cultural and human, material, and environmental contexts. The more assumptions designers make about their stakeholders themselves, the context the stakeholders work and live within, and the stakeholders' needs, the more likely it is that designers will make mistakes and come up with the right solution to the wrong problem. Only by gathering information to interrogate those assumptions can designers make informed decisions about what is important to stakeholders. The evidence-based requirements and constraints generated will then lead to better problem statements and ultimately more desirable final design proposals.

SELECTED EXERCISES

Exercise 8.1

When students have been given a design project that involves changes or modifications to existing spaces, such as classrooms, have the students visit a variety of classrooms around campus with an eye toward the differences in the spaces that impact any design solution or

create constraints that may not have been considered. The students can be guided in the review by providing them with a list of suggested classrooms to visit to show a variety of room arrangements, available wall space, seating arrangements, and number of exits/entrances to the room. Once students have completed this review, have them share with the class what they learned, particularly as it may impact any designs being considered.

Exercise 8.2

Using Table 8.1 as a starter, create a worksheet for students with a column added to the right. In this additional column, have the students fill in the specific information need for designing a playground or the design project being used in class, trying to find at least one specific source for each type of information. Use the information gathered by the students as a starting point for a class-wide discussion so that everyone is involved in thinking about where different types of information can be found.

Exercise 8.3

Create an incomplete version of Figure 8.2, the how-why diagram, using a problem new to the students. Fill in the selected design and the options below it in the diagram, and leave the additional options for solving the problem blank. Have the students work in teams to come up with other options for solving the problem. Have teams share with the rest of the class. Guide the conversation to ensure that the new ideas focus on the *why* behind the problem to be solved, rather than jumping to a potential solution.

ACKNOWLEDGMENTS

The Snooze-o-Matic was a witty idea originally proposed in an issue of *MAD Magazine*. The how-why example in Figure 8.2, created by David Radcliffe, was developed around this concept.

REFERENCES

Atman, C. J., Adams, R. S., Cardella, M. E., Turns, J., Mosborg, S., & Saleem, J. (2007). Engineering design processes: A comparison of students and expert practitioners. *Journal of Engineering Education, 96*(4), 359–379. http://dx.doi.org/10.1002/j.2168-9830.2007.tb00945.x

Atman, C. J., Chimka, J. R., Bursic, K. M., & Nachtmann, H. L. (1999). A comparison of freshman and senior engineering design processes. *Design Studies, 20*(2), 131–152. http://dx.doi.org/10.1016/S0142-694X(98)00031-3

Crismond, D. P., & Adams, R. S. (2012). The informed design teaching and learning matrix. *Journal of Engineering Education, 101*(4), 738–797. http://dx.doi.org/10.1002/j.2168-9830.2012.tb01127.x

Kilgore, D., Atman, C. J., Yasuhara, K., Barker, T. J., & Morozov, A. (2007). Considering context: A study of first-year engineering students. *Journal of Engineering Education, 96*(4), 321–334. http://dx.doi.org/10.1002/j.2168-9830.2007.tb00942.x

Leckie, G. J., Pettigrew, K. E., & Sylvain, C. (1996). Modeling the information seeking of professionals: A general model derived from research on engineers, health care professionals, and lawyers. *The Library Quarterly: Information, Community, Policy, 66*(2), 161–193. http://dx.doi.org/10.1086/602864

Preece, J., Rogers, Y., & Sharp, H. (2002). *Interaction design: Beyond human-computer interaction.* New York, NY: Wiley.

Rosson, M. B., & Carroll, J. (2001). *Usability engineering: scenario-based development of human-computer interaction.* San Francisco, CA: Morgan Kaufmann.

Stone, D., Jarrett, C., Woodroffe, M., & Minocha, S. (2005). *User interface design and evaluation.* San Francisco, CA: Morgan Kaufmann.

Wertz, R. E. H., Fosmire, M. J., Purzer, S., & Cardella, M. E. (in press). Assessing information literacy skills demonstrated in an engineering design task. *Journal of Engineering Education.*

Clarify the Task

CHAPTER 9

MAKE IT SAFE AND LEGAL

Meeting Broader Community Expectations

Bonnie Osif, The Pennsylvania State University

Learning Objectives

So that you can make students aware of their obligations as professional engineers, upon reading this chapter you should be able to

- Describe the concept of inherent safety and its implications for information across the life cycle of a new product or system

- Distinguish between a specification, a standard, and a regulation in the context of engineering design

- Locate and obtain relevant standards and regulations pertinent to your design project

INTRODUCTION

In addition to understanding user needs and contextual factors, the design team needs to consider issues of safety, legal constraints, and/or professional standards for performance or interoperability. These matters need to be addressed early on in the design process as part of clarifying the task. Safety is a paramount consideration that begins at the outset of a design project and which spans the entire life cycle of any product, process, or system. If the design team fails to take into account the need for certification to meet a required standard for safe use or issues of compatibility with other systems, then the design effort may be wasted. It is best to understand such design constraints and opportunities early in the design cycle. While a client may or may not know the relevant professional standards and regulations, the design team needs to be aware of them so that their solution is safe and legal.

Design failures can arise from very simple assumptions that are made early in a project, from issues that are taken for granted or are so obvious that no one thinks to ask or to check. On September 23, 1999, the Mars Climate Orbiter entered the Martian atmosphere rather than its planned higher orbit and was destroyed. During the flight to Mars, NASA engineers tried unsuccessfully to correct the trajectory. The failure was due to a very simple error—NASA planned the mission acceleration in metric units, while the builder of the orbiter used English units. As NASA explained, "The 'root cause' of the loss of the spacecraft was the failed translation of English units into metric units in a segment of ground-based, navigation-related mission software" (Isabell & Savage, 1999, para. 6). This failure was due to a very simple error, but the cost was high in dollars, effort, and prestige.

Clarity of information can avoid costly and sometimes deadly errors in engineering. As with the Orbiter, errors can be as basic as a unit of measurement mismatch or as complex as selecting the wrong materials for a particular environment. To address the need for clarity and specificity in engineering, a number of standards have been developed by various organizations with the goal of addressing conformity, reliability, compatibility, and safety. These include specifications, standards, codes, and regulations. While all share some commonalities, there are distinct differences among them—who creates and authorizes them, if they are mandatory or voluntary, and how they are promulgated—but in general they have some very important commonalities, such as providing guidance for the engineer to meet at least minimal levels of safety, structural integrity, and physical limits, among other requirements. In our increasingly global economy, they also provide the engineer and the consumer with some information on what level of confidence to place in a design solution. For example, the Institute of Electrical and Electronics Engineers (IEEE) Standard 802.15 for wireless communication assures buyers of cell phones that, regardless of the location of manufacture or the name on the case, the phone will operate as they expect, wherever they are in the world (IEEE, 2002).

COMMON CHALLENGES FOR STUDENTS

Students often underestimate the centrality of safety in the design process and do not take into account relevant codes, regulations, and standards in the choices they make in designing a product. Examples of these challenges include the following:

- Considering safety as an integral part of the design process.
- Explaining how well-documented design specifications can improve safety in design.
- Finding, reading, interpreting, and applying the relevant information from standards, codes, and regulations with completeness, precision, and accuracy.
- Thinking globally rather than provincially when considering relevant standards and regulations.
- Considering concepts of standard sizes and interchangeability in components.
- Specifying with precision the composition and performance of materials, especially under different operating conditions.

SAFETY

Safety considers the avoidance, prevention, and diminishment of hazards and their potential impact on people and things. Most safety issues involve ensuring that a source of energy does not come in contact with a person or piece of equipment in an uncontrolled manner such as to cause injury or damage. The design hierarchy for ensuring safety is to (1) separate the energy source from the person or place where it can do damage; (2) reduce, restrict, or eliminate possible pathways for the energy to reach the person or place; and (3) as a last line of defense, protect the person or place from damage from the energy.

Safety should be designed in from the very beginning of a project rather than being added on at the end. In the context of chemical engineering, Trevor Kletz subscribes to the notion that "what you don't have, can't leak" (Mannan,

2012, para. 1). The basic idea is that the design solution should be one that is inherently safe even if something does go wrong. Reduction or elimination of hazards is the goal. Tragic instances of explosions, injuries, and death indicate the need for designing to minimize the hazardous materials and processes in a plant. While it is impossible to completely eliminate accidents, they can be decreased by limiting the amounts of hazardous materials used, substituting safer materials, simplifying design, and designing for projected worst cases. For example, limiting the amount of a caustic agent present, using a less caustic agent, moving the agent to a safer location, and constructing a containment system are examples of designing for inherent safety. While the concept is integral to chemical engineering, it can be applied to every engineering discipline.

To achieve inherent safety, it is critical to remember that safety should be a primary consideration at all stages of the product life cycle, from needs assessment, through design and manufacture to the use of the product, and ultimately to its disposal at the end of its useful life. It is a factor in concept development, selection of materials, detailed design of equipment and processes, design of training, and work conditions, and it must consider all people who might come in contact with the product at every stage of its life cycle. This includes the people who make it, move it, install it, operate or use it, maintain it, and repurpose or recycle it. Consideration of safety issues saves time and money in the long run, and avoids subsequent problems with product recalls and related legal issues.

Of increasing importance is safety after the useful life of a product is over. Safe, efficient recycling or disposal is a critical design consideration. While for some products and some countries or jurisdictions this is merely a desirable outcome, for other countries it is mandatory and enforced by law. All engineers should consider such life cycle safety within both the legal framework of the market and the ethical framework of the profession.

Inherent safety is the foundation upon which good engineering design rests. The most innovative product is worthless if the process to make it is dangerous or if its use is high risk for the consumer. The most efficient manufacturing is pointless if it cannot be done without harm to the workers. Understanding the basic aspects of safety and the appropriate standards and regulations—whether customized for a local facility or regulated locally, nationally, or internationally—is essential.

The question remains, what is safe or what is safe enough? According to Vesiland and Gunn (2011), "the key principle is that the level of safety be understood and fully communicated to the user, and that any deviance from this accepted level of safety without full understanding of the user is unethical conduct" (p. 162). For example, bungee jumping from a bridge into a rocky gorge is not advised for people who have particular medical conditions. However, if the jumping facility has been designed properly with adequate clearance from the platform; has equipment that meets all the appropriate standards for manufacturing, installation, and inspection; and has properly trained staff, there is a level of trust in the safety of the activity. Safety is about reducing the likelihood of something going wrong and the severity of the consequences to people or property if something does go wrong.

Safety in design applies to the things we use in our everyday lives—from the coffeepot we plug in for breakfast to the alarm clock we set at night—as much as it does to large, complex engineering projects.

DESIGN SPECIFICATIONS

A design specification describes a product or system in terms of what it is capable of doing, by using both a metric and a value (Haik & Shahin, 2011). In contrast to a design requirement (see Chapter 7), which focuses on needs or desires, the specification is a statement of expected performance. For example, "Product A will lift x number of pounds y feet in z seconds." With this precise information, the designer can begin to plan the development of the product. However, this is not a once and done process. As the product or system is developed and tested, new information is discovered and must be accommodated in the design specifications. Specifications often will be adjusted or refined as the design process develops and actual constraints and costs indicate that some specifications must be reconsidered. As an example, it is discovered that although Product A can easily lift the specified number of pounds the specified number of feet, doing so at the rate determined in the original specification would cause damage to the merchandise. Therefore, the rate needs to be adjusted, and the final specification would reflect that change.

The benefits of specifications are many, especially if careful documentation is kept of each aspect of the design process, including who is responsible and when the various aspects have been accounted for or changed. This itemization and accountability may limit errors, inefficiencies, and poor communication, especially of important changes. It also helps focus attention on specification targets and inclusion of individuals such as safety specialists, and it tracks progress on the project. For many projects a formal, written checklist is recommended, although there are instances when a less formal process is acceptable.

Analysis of the design specifications of previous products or systems can reduce risks and increase safety by carrying knowledge from past projects forward so that mistakes are not repeated. For example, if a pedestrian bridge is being designed, it is useful to know what issues and solutions have worked and what problems have been noted in the past. If specifications include a particular appearance and materials that have been known to cause problems in the past, it would be beneficial to already know about, for example, the wobbly bridge problem and alter the specifications to adjust for the vibration issues with dampers and vibration absorbers (Hales & Gooch, 2004).

STANDARDS

Standards are consensus documents that consolidate knowledge and best practices aimed at improving safety, reliability, quality, efficiency, interchangeability, and testing, and creating a consistent measurement, terminology, and use of symbols (see, e.g., de Vries, 1999). They are written by a group of subject matter experts and many are updated frequently, particularly after a problem or failure has been noted. Standards can apply to one specific company or to an entire industry. They can be created by local or national government groups, a collection of countries such as the European Union, or by nongovernmental organizations or professional societies. While adhering to standards is voluntary, it is good practice to take into account the standards that are relevant to both the location where the product or system designed will be used and the relevant professional organizations of the specific area of engineering. Box 9.1 contains the American Society of Mechanical Engineers (ASME) definition of a standard.

> **BOX 9.1**
>
> **American Society of Mechanical Engineers (ASME) Definition of a Standard**
>
> A standard can be defined as a set of technical definitions and guidelines—"how to" instructions for designers, manufacturers, and users. Standards promote safety, reliability, productivity, and efficiency in almost every industry that relies on engineering components or equipment. Standards can run from a few paragraphs to hundreds of pages and are written by experts with knowledge and expertise in a particular field who sit on many committees.
>
> Standards are considered voluntary because they serve as guidelines, but they do not of themselves have the force of law. ASME cannot force any manufacturer, inspector, or installer to follow ASME standards. Their use is voluntary.
>
> Standards become mandatory when they have been incorporated into a business contract or incorporated into regulations.

There are standards that use the term *specification* or *spec*. These are different from design specifications and are usually interchangeable with those called *standards*. One of the most widely used are the Military Standards (MIL SPECs), which are standards set by the United States military for both engineering and nonengineering requirements.

Standards are a major source of information for designers, providing a look at best practices and successful design processes. Reviewing standards allows designers to benefit from the wisdom and experience of others, rather than reinvent the wheel each time. This results in time and money savings and the avoidance of unsuccessful or inefficient processes. Engineering, like so many other fields of endeavor, benefits from the accumulated wisdom of previous practitioners, and standards are a formal way of documenting those advances. Standards also allow for increased interchangeability and interoperability. For example, parts, tools, and training can be consistent across a system if the same standard is used for a product. Travelers are well aware of the variety of electrical plugs used in different countries and the need for bringing adaptors. Until recently most chargers were specific to each brand of cell phone, requiring the purchase of a new charger every time one bought a new phone. The move by many manufacturers to the USB standard has changed that.

An important source of information about standards can be found on the National Institute of Standards and Technology website (http://www.nist.gov/director/sco/index.cfm). The site has a number of useful links and an interactive map to check standards from around the world, including regulations, relevant news, and much more, including links to standards creators and providers.

The importance of standards is clear from the statement from the American Society of Civil Engineers, which states that "all engineering graduates should have at least a rudimentary knowledge of the standards system and standards development, standards as they affect engineering design and practice in general and some knowledge of standards specific to their specialized field" (Kelly, 2008, p. 159). The importance of standards cannot be overemphasized in the design process. They affect every aspect of our lives and bleed over into the popular media. News reports frequently document the tragic results of nonadherence to existing standards or the need for revised standards. Examples include poorly designed cribs with slats or spindles too far apart, toys containing lead, toys with parts that can cause choking, flammable clothing, and unsafe drug manufacture. Adherence to relevant standards and

review and updating of existing standards is a critical engineering practice. Standards impact almost every aspect of our lives, from toy safety to strength of materials in airplane cockpits to materials used in medical procedures. While many standards can be searched in specific databases such as ASTM (American Society for Testing and Materials; http://www.astm.org) or IEEE Xplore (http://ieeexplore.ieee.org/xpl/standards.jsp), both commonly accessible in full text at academic libraries, a more general subject search can be done in the NSSN standards database (www.nssn.org), provided by the American National Standards Institute (ANSI). This resource searches U.S. and international standards from a wide range of sources and provides access information.

Finding appropriate standards can be a difficult task. While NSSN is an excellent source, students frequently have trouble discovering the correct terminology to search. For example, knowing that there is a standard used in the production of the Lego building block toy does not make it easy to find the ASTM standard, "Standard Consumer Safety Specification for Toy Safety" (ASTM F963). Local documentation, stated requirements from the customer, and utilization of a knowledgeable person to review the appropriate standard resources will help ease the process of locating the correct standard.

CODES AND REGULATIONS

The term *code* is commonly used interchangeably with the term *standards*, although there is a definite distinction between the two terms. ASME notes that "a code is a standard that has been adopted by one or more governmental bodies and has the force of law" (ASME, 2012, "What is a code?"). Examples are the ASME Boiler and Pressure Vessel Code, International

Building Code, the National Fire Protection Association's Fire Code (NFPA 1), and the National Electrical Code, among others. Adherence to the appropriate code is critical. Codes provide a level of dependability and reliability with wide acceptance. A product that meets or exceeds code specifications provides important information to those using or affected by the product. For example, the ASME Boiler and Pressure Vessel Code

> establishes rules of safety—relating only to pressure integrity—governing the design, fabrication, and inspection of boilers and pressure vessels, and nuclear power plant components during construction. The objective of the rules is to provide a margin for deterioration in service. Advancements in design and material and the evidence of experience are constantly being added. (ASME, 2013, "About the Code")

Utilization of this type of code provides a level of exactness and trustworthiness that is recognized, often internationally. The result of not adhering to codes can be fines, increased inspections, radical renovations, and lost business.

Regulations are the laws that require the adherence of a product to codes or other technical requirements. They ensure the health and safety of the product with consideration of consumer safety, environmental impact, and user safety, among other aspects, and are frequently based on standards. U.S. regulations are recorded in the Code of Federal Regulations (CFR). Regulations from other countries can often be found on the Library of Congress' Global & Comparative Law Resources website (http://www.loc.gov/law/find/global.php). The website link to the Guide to Law Online (http://www.loc.gov/law/help/guide.php) can be especially useful. However, finding the appropriate regulation might be dif-

> **BOX 9.2**
> **U.S. Government Websites for Regulations**
> *LexisNexis State Capital (fee database)*
> http://www.lexisnexis.com/en-us/products/lexisnexis-state-capital.page
> *NIST Regulations*
> http://www.nist.gov/standardsgov/regulations.cfm
> *Office of Information and Regulatory Affairs*
> http://reginfo.gov
> *Federal Register (1994–current)*
> http://www.gpo.gov/fdsys/browse/collection.action?collectionCode=FR
> *Code of Federal Regulations (1996–current)*
> http://www.gpo.gov/fdsys/browse/collectionCfr.action?collectionCode=CFR
> *Regulations.gov*
> http://Regulations.gov

Clarify the Task

ficult, or it may not be included in this resource. In that case, it is best to search the U.S. government websites for laws and regulations that might impact the design project (see Box 9.2).

INTERNATIONAL ISSUES

Designers need to know the market or markets that will use the product or system being designed, as the standards vary from jurisdiction to jurisdiction. While there are still standards unique to a particular country, increasingly standards are shared within cooperating groups of countries, such as the European Union. Major international standards organizations include the International Organization for Standardization (ISO), the International Electrotechnical Commission (IEC), and the In-

> **BOX 9.3**
>
> **Sources of Standards Information**
>
> **Government provider:**
>
> **NIST Global Standards** (provides links to a number of resources)
>
> > http://gsi.nist.gov/global/
> > index.cfm/L1-5/L2-44/A-171
>
> **Commercial providers:**
>
> **Document Center**
>
> > http://www.document-center.com
>
> **IHS Standards Store**
>
> > http://global.ihs.com
>
> **SAI Global**
>
> > http://www.saiglobal.com
>
> **Techstreet Store**
>
> > http://www.techstreet.com

ternational Telecommunications Union (ITU). Also, there are a number of other organizations that focus on very specific areas, such as timber, aluminum, or illumination.

Whether a product designed by students is to be used internationally or if it is specifically for a given country, as is becoming common in service learning courses, attention must be paid to the standards and regulations that exist in the relevant market. A number of companies provide access to standards (see Box 9.3); however, there are instances in which the only way to obtain the relevant standard is to contact the appropriate government office directly, which can be a slow process.

LOCATING AND ACESSING STANDARDS

Identification of and access to the standards and regulations for student projects can take a number of paths (see Box 9.4 for examples). It may be as easy as consulting a list of databases subscribed to by the university's engineering library and conducting a subject search to obtain a downloadable copy of the appropriate full text standard. ASTM and IEEE Xplore are commonly held by most engineering libraries. In other cases it might entail a search of the catalog to find the call number of a print standard. Often a student will be searching by subject and either not know or not be concerned about the specific sponsoring organization. In this case the NSSN standards database (www.nssn.org) might be the best place to start the search, then once the standard is identified the library's catalog and databases can be consulted to determine whether a document is accessible. When a standard is not available locally, it can usually be obtained in minimal time via either interlibrary loan or a purchase request. The exception is for countries whose standards are not available from the major standard provid-

> **BOX 9.4**
>
> **Standards Websites**
>
> **ASTM International**
>
> > http://www.astm.org
>
> **IEEE Xplore Digital Library—"Standards"**
>
> > http://ieeexplore.ieee.org/xpl/
> > standards.jsp
>
> **National Institute of Standards and Technology**
>
> > http://www.nist.gov
>
> **NSSN Search Engine for Standards**
>
> > http://NSSN.org
>
> **The Society for Standards Professionals— "National Standards Bodies"**
>
> > http://www.ses-standards.org/
> > displaycommon.
> > cfm?an=1&subarticlenbr=54
>
> **Standards.gov**
>
> > http://standards.gov/
>
> **World Standards Services Network**
>
> > http://www.wssn.net/WSSN/
> > index.html

REALITY CHECK 9.1

A class has been assigned to design playground equipment for a local park. The trustees of the park provided a list of requirements that include the types of equipment that they want and the age range of the children who will be using the park. With this information, the class needed to devise usable specifications for the requested equipment. Using the weight and height information from the Center for Disease Control growth charts (www.cdc.gov/growthcharts), the students created a specification for the weight and height and other pertinent physical parameters of the children for the various equipment to address the appropriate age groups. Searching the ASTM standards they then located appropriate national standards for playground equipment from the Consumer Product Safety Commission's Public Playground Safety Handbook (http://www.cpsc.gov/CPSCPUB/PUBS/325.pdf). State and local standards and regulations were then reviewed for the specific locale of the playground. Finally, the Americans with Disabilities Act of 1990 (http://www.ada.gov/2010ADAstandards_index.htm) was consulted to determine what specific accessibility issues needed to be addressed.

ers. In that case the best path is to use the SES— The Society for Standards Professionals website (http://www.ses-standards.org) and go directly to the country in question. Comparing standards on a particular topic is also a very good exercise for students to increase their understanding of the spectrum of expectations around the globe.

SUMMARY

While many aspects of safety are addressed in the standards, codes, and regulations, best practices and local knowledge all need to be considered as well. Safety is a critical aspect of all design and must be considered as integral at every level of the process. It is

doubtful if any combination of standards and regulations can comprehensively address every aspect of the product or system being designed—its processes, location, and personnel—so other safety features must be incorporated into the design process. Documentation is important to memorialize the steps taken for increased safety, to inform those that follow, and to serve as an evolving template for future safety improvements. Safety builds on industry standards as well as local, learned knowledge.

Differentiating codes, standards, and specifications can be challenging. Understanding which are mandatory by law (regulations), what is mandated by customer (specifications), and what is voluntary but worthy of serious consideration (standards) can be a difficult task, and students need to practice thinking about the roles of regulations, specifications, and standards in their design projects.

By incorporating user needs (Chapter 7), context (Chapter 8), and best practices of the profession, students will create a much more robust problem statement that will help frame the potential solutions they will generate, using techniques discussed in the following chapter, and evaluate those solutions, as will be discussed in Chapter 11.

SELECTED EXERCISES

Exercise 9.1

Your students have been asked to design a waste disposal system for a rural village in Haiti devastated by the 2010 earthquake. What physical and financial issues will they need to address? What standards are relevant to this project from both the Haitian government and from professional standards governing this field of engineering?

Exercise 9.2

Failures can be instructive. Have students review of one of the following cases to stimulate discussion of the role of standards and regulations and their limitations. Discussion questions may include the following: Were standards followed? Were the standards adequate? How could the standards be changed? Have the standards been changed? What has been learned? Suggested topics include the following:

- Breach of the flood control system in Louisiana after Hurricane Isaac in 2012
- The London Millennium Footbridge (opened and closed in June 2000; reopened in 2002)
- Metal hip replacement implants
- Video recorders (VHS versus Betamax)

Exercise 9.3

Consider the scenario where students are designing a large-scale food dryer. They plan to use local materials and are seriously considering plastic piping. Have students investigate whether there are standards for the materials they can use, since the materials will be in direct contact with the food in the particular country in which they will be working.

REFERENCES

American Society of Mechanical Engineers. (2013). *Boiler and pressure vessel code.* New York: ASME, Inc. Retrieved from https://www.asme.org/shop/standards/new-releases/boiler-pressure-vessel-code-2013.

American Society of Mechanical Engineers. (2012). *Standards & certification FAQ.* New York: ASME, Inc. Retrieved from http://asme.org/kb/standards/about-codes-standards

De Vries, H. J. (1999). *Standardization: A business approach to the role of national standardization organizations.* Boston: Kluwer Academic.

Haik, Y., & Shahin, T. M. (2011). *Engineering design process* (2nd ed.). Stamford, CT: Cengage.

Hales, C., & Gooch, S. (2004). *Managing engineering design* (2nd ed.). London: Springer-Verlag. http://dx.doi.org/10.1007/978-0-85729-394-7

Institute of Electrical and Electronics Engineers. (2002, June). *802.15.1-2002—IEEE standard for telecommunications and information exchange between systems—LAN/MAN—Specific requirements—Part 15: Wireless medium access control (MAC) and physical layer (PHY) specifications for wireless personal area networks (WPANs).* Retrieved from http://dx.doi.org/10.1109/IEEESTD.2002.93621

Isabell, D., & Savage, D. (1999). *Mars Climate Orbiter Failure Board releases report, numerous NASA actions underway in response.* Release: 99-134. Washington, DC: NASA. Retrieved from http://www.nasa.gov/home/hqnews/1999/99-134.txt

Kelly, W. E. (2008). Standards in civil engineering design education. *Journal of Professional Issues in Engineering Education and Practice, 134*(1), 59–66. http://dx.doi.org/10.1061/(ASCE)1052-3928(2008)134:1(59)

Mannan, M. S. (2012). Trevor Kletz's impact on process safety and a plea for good science—An academic and research perspective. *Process Safety and Environmental Protection, 90*(5), 343–348. http://dx.doi.org/10.1016/j.psep.2012.06.006

Vesiland, P. A., & Gunn, A. S. (2011). *Hold paramount: The engineer's responsibility to society.* South Melbourne, Victoria, Australia: Cengage Learning Australia & New Zealand.

CHAPTER 10

DRAW ON EXISTING KNOWLEDGE

Taking Advantage of Prior Art

Jim Clarke, Miami University

Learning Objectives

So that you can encourage students to explore a wide variety of potential solutions before committing to a particular course of action, upon reading this chapter you should be able to

- Define and understand the purpose of examining prior art
- Identify a variety of technical information sources of prior art
- List tips and strategies for searching scholarly and popular technical literature
- Utilize team processes for examining and applying prior art effectively

INTRODUCTION

Once a student design team has thoroughly explored the specific needs of the project stakeholders and the safety and performance constraints the team needs to meet, design team members start to formulate potential solutions. At this point, it is important for students to cast the widest net of possible solutions. In addition to using traditional intra-team techniques such as brainstorming, students need to look outside the immediate knowledge of the team and investigate how others have solved similar problems, an activity that is often referred to as *investigating prior art*. The investigation or study of prior art is a vital part of the design process because it encourages designers to discover and consider as many options as possible before they begin the process of choosing their own solution. Designers then have a decisive advantage for success because they will have gained an awareness of all the prevalent solutions in the market, not just the ones they might have been familiar with before the assignment. Once information is gathered and synthesized from prior art, designers can proceed with a comprehensive benchmarking process to choose the best solution possible for their project (see Chapter 11).

When design teams study prior art, they are essentially learning the state of the art related to their project. This understanding is gained through the systematic gathering of technical literature. To conduct a far-reaching literature search, undergraduate design teams explore all aspects of business and engineering literature collections. Books (monographs and series), encyclopedias, scholarly journal articles, conference papers, dissertations, patents, and standards are common information resources utilized by designers. Design projects are often related to consumer products or capital goods, so invaluable information may be accessed from material produced by and about corporations, such as press releases, product manuals, annual reports, trade publications, and industry blogs. Marketing collateral such as brochures, sales sheets, and catalogs may also provide useful technical information. Successful design teams collect and review as much relevant information as possible as they investigate the prior art.

COMMON CHALLENGES FOR STUDENTS

A key challenge for student design teams involves maintaining a proper attitude toward searching prior art. For example, in a typical senior design class, it is only natural for students to feel confident in and want to demonstrate the knowledge and skills they have gained through their classes and labs. Thus, engineering students frequently want to build solutions from first principles, rather than building on solutions or technologies that already exist. There is also a common tendency for design teams to choose a solution before they even start investigating the prior art, what is commonly known as *design fixation* (Dahl & Moreau, 2010). The team wants to jump into the solution without really embracing the problem, and as a result, they may get far along the path of prototyping a solution before they realize there might be a fundamental flaw in their approach, or another cheaper, more effective approach. The cost of changing approaches is much higher the farther along the design process one goes, so exploring the breadth of solutions up front is essential to save time and money and to ensure optimal performance of the artifact.

Building a Stair-Friendly Stretcher

Searching the prior art can lead to unexpected discoveries that can become decisive advantages. For example, an emergency medical services employee served as a capstone project stakeholder by inviting the students down to the municipal firehouse, where they viewed a foldable, chair-like stretcher used by EMS workers to transport patients up or down staircases as they proceed to the EMS vehicle outside. The students learned that EMS workers are always at risk of hurting their backs during the transportation process, and that the straps on the staircase stretchers are not adequate for restraining patients for their safety. As a consequence, the student team was tasked with developing a motorized staircase stretcher with improved restraints that would fit into an EMS vehicle properly. Another requirement of the design project involved designing a removable motor in the case of a breakdown.

As the student design team conducted background research, a key question that emerged involved their curiosity about why a motorized staircase stretcher had not already been introduced into the marketplace by one of the product manufacturers. A general search through an ordinary Web browser led the student team to a firefighter/EMS blog that contained a press release for a company called Paramed Systems located in Utah that had developed a motorized staircase stretcher. The students became disheartened, but their engineering librarian encouraged the students to learn why it had not yet emerged as a significant product in the marketplace. The librarian also encouraged the students to learn about how the Paramed Systems product was constructed.

The effort of conducting a quick inventor/assignee patent database query with the name of the Paramed Systems chief executive officer led the students to the actual motorized product patent that could explain all of the product details. Another simple search for the company's name on the website www.youtube.com revealed a conventional demonstration video in which a company representative explained key facts like the heavy weight of the product, the high price of the product, and its un-removable motor. The student design team was then able to use all of the information about the competitor product to their advantage as they developed a solution more appropriate for the project stakeholder. Searching the prior art thoroughly empowered the capstone team to continue in the design cycle process with great success.

One reason student designers are susceptible to this mindset is that traditional undergraduate engineering curricula focus on working textbook problems rather than on open-ended, more authentic problem solving. Literature searching is often regarded as a soft skill, and engineering faculty rarely focus much class time preparing students to gather information before the capstone experience. Undergraduate engineering students may have examined some technical literature during their first three years of course work, but that is often the exception rather than the rule. The probability that students will instinctively place a higher value on technical literature research at the outset of their capstone course is also doubtful, if it

has not been reinforced throughout the engineering curriculum. As a consequence, there is always a high risk that undergraduate design teams come into a course considering prior art research as a low priority.

Another key challenge student designers will face as they search prior art involves the time constraints related to capstone and other types of design projects. In many cases, capstone design projects must be completed during the course of only one or two semesters. Immediate pressure for progress exists at the outset of all capstone design projects, and unexpected delays in identifying stakeholder needs may compromise the start of the literature search. The student design team advisors also face

Synthesize Possibilities

pressure to make certain their teams progress steadily toward producing a final artifact. For all of these reasons, time management is a vital task for design teams as they explore the prior art, and instructors need to emphasize the fact that time spent searching the literature up front will be as useful, or more so, as time spent in the lab constructing the final artifact.

Young engineers need to avoid these common pitfalls by maintaining a practical attitude toward the benefits they can receive from all of the available and relevant information resources. The careful study of prior art will help students proceed along the most promising path for a good solution. It will also provide documentation to help persuade stakeholders that the students' design solution is based on the best practices approach to the problem (see Chapter 13 for more about communication with stakeholders). With strong information skills gained from this experience, students will also be more attractive to employers and confident in their ability to be lifelong learners (Strouse & Pollock, 2009).

TECHNIQUES AND TOOLS FOR EFFECTIVE INFORMATION GATHERING

The main focus of synthesizing solutions is to generate the broadest selection of potential solutions to the design problem. For example, students need to be thinking about ways to cross a river rather than how to build a bridge in this phase. This type of thinking opens up the design space to allow for a much richer set of solutions that might include ferries, kayaks, zip lines, stepping stones, and so forth instead of just different styles of bridges. Not all ideas will be practical or even desirable, but transformative products come from thinking outside the box. The key is for students to not become self-conscious about providing ideas—thus the common mantra *there are no bad ideas* when brainstorming. Much has been written about ideation and brainstorming techniques, with IDEO (Kelley, 2001) being a current model for best practices, and Frog Design's (2013) Collective Action Toolkit providing activities to spur innovation and action at the community level.

When design teams are ready to begin the process of searching the prior art, they should adopt a systematic approach for determining what kinds of information they ought to gather. Techniques can be used to generate concepts and ideas. Attribute listing involves separating a problem into smaller elements and addressing each one separately (Morgan, 1993). Case-based reasoning involves the study of old designs to inspire new ones (Kolodner, 1993). Lateral thinking involves developing a radical statement about a problem or potential solutions to challenge designers to consider more diverse ideas (De Bono, 2009). Group brainstorming is a popular technique for capstone teams to generate a large quantity of creative and diverse ideas regardless of whether or not all of them may be used to solve a given problem (Wang, Cosley, & Fussell, 2010).

To make brainstorming systematic for groups, card-based tools are sometimes used to organize and focus the process. A good example of a card-based tool that might be worth trying is called an *ideation deck*. This method is distinctive among other card-based tools because it includes specific parameters directly related to a design problem. A team starts an ideation deck by clearly defining the design challenge in writing. Then the team must define a minimum of three factors most relevant to the design project. These factors can be abstract or specific. These three factors are then known as *category suits*. A list of specific examples for these factors must be generated and used to make *instance cards* for each category suit. Then the team collaborates to develop content for the instance cards. Once content is established for the instance cards, the back of the cards can be color coded based on suit. At this point the ideation deck is now complete, and cards can be laid out in a grid that intermixes the instance cards. The design team can then discuss card combinations within specific categories and discover provocative options to consider. An exercise like this can help to improve creative thinking that will then expand the search through prior art (Golembewski & Selby, 2010).

Other examples of ideation techniques include Wodehouse and Ion's (2012) ICR (inform, create, reflect) Grid method, which requires designers to find a piece of information, usually an image, in an Internet search and pass it on to the next designer, who applies it to the design problem. In their study, the approach led to more novel and detailed solutions than the non–information integrated approach, and they also found that information literacy instruction, not just familiarity with Internet searching, was important in sourcing high-quality information, leading to more robust solutions. IDEO's Tech Box (Kelley, 2001), which is filled with technologies that designers can manipulate during ideation, similarly provides external sources of inspiration and the ability to make new connections from existing artifacts.

While information can be integrated using the simple methods mentioned, there is also value in conducting dedicated searches for potential solutions. Relying only on their prior knowledge can leave large holes in the solution space investigated by students. For example, when looking for water purification solutions for a remote village, if the students are only aware of natural percolation techniques, they will have missed out on all the distillation and disinfection options that might be much more cost-effective and efficient for the situation they are working with. Having students conduct a systematic survey of the current state of technology will avoid gaps in their analysis that can lead to uncomfortable questions in the students' ultimate design presentation.

When carrying out such a search, even with a proper attitude and strong time management skills, novice designers face the challenge of quickly becoming efficient users of literature collections. As soon as design teams have a clear understanding of stakeholder needs, they should refresh their knowledge about the breadth of their institution's literature collection and how to efficiently find information with online catalogs, subject guides, indices, and literature databases. Some universities provide library instruction seminars near the start of new capstone courses to refresh and update student awareness of the available technical literature collection. Other courses have designated embedded librarians who are available for consultation during class time or at appointed times outside of class. Design teams should take advantage of these resources to make the best use of their limited time. Even if library instruction sessions are not made available, design teams should establish a working relationship with engineering librarians right away. Subject librarians are often few in numbers even at the largest technical universities, so design teams need to start early in scheduling initial meetings and establishing collaboration.

When initial meetings do occur, design teams need to be prepared to thoroughly explain the project task to engineering librarians, including the team's initial thoughts about what information they already know

Synthesize Possibilities

Design Information Audit

This worksheet requires you to take an inventory of the information you currently have and information you still need to acquire in order to successfully complete your design project. You may receive all the information you think you need from the client you are working with, but searching for best practices and alternative solutions that others have produced may enrich your final project. You can use a separate page for each information facet below.

Overall Project Statement

(one sentence synopsis of project goal) _____

	Information Already Known	Evidence (i.e., Sources)	Additional Information Needed	Proposed Sources of Needed Information
Stakeholder Needs What does the client want? What does the user need? How do you know?				
Foundational Information What kind of task is it? What are the basic principles for solving this type of problem? What do you need to optimize? How do you measure performance?				
Best Practices What have other people done to solve this type of problem?				
Materials/Components What materials are you planning on using? What properties are important for the particular use? What properties does your material have?				
Regulations and Standards What legal requirements or industry standards do you need to adhere to in your solution?				
Intellectual Property Are you inventing something, or using the inventions of others to solve this problem?				

FIGURE 10.1 Design information audit. (Courtesy of Michael Fosmire.)

and what they still need to find out about their project task (see Figure 10.1). After conducting a reference interview, engineering librarians will provide some practical instruction about how to access the technical literature collection with database and catalog query demonstrations. All literature databases and indices have distinctive features, but Boolean logic, key words, date range control, controlled vocabulary, truncation, and search histories are examples of universal query elements that can be used with most online literature searching tools. Engineering librarians can help students identify the most relevant online tools and can demonstrate specific query tactics for effective use. Design teams must be responsible for conducting their own literature searches and be prepared for the possibility that their literature searching process will last a significant period of time. In some instances, searching, understanding, and integrating prior art for a capstone design project may require the majority of a semester to complete, and some institutions have a pre-design course that focuses on problem definition and prior art searching, with the formal capstone design course focused on the build portion of the design process. No matter the amount of time required for any specific design project's literature search, design teams should always consult with engineering librarians at least a few times during the process. Engineering librarians can offer invaluable suggestions to improve queries and

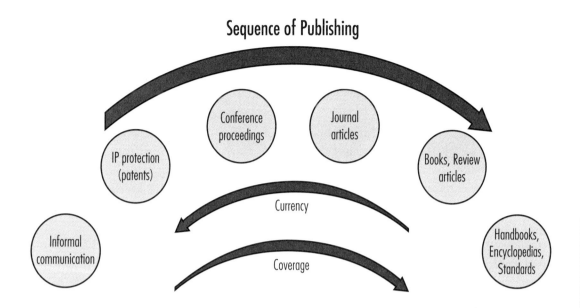

FIGURE 10.2   Characteristics of technical information.

identify resources designers may not have yet considered.

The quantity and types of technical literature required for specific design projects will always vary, but design teams should take it upon themselves to look at all types of engineering literature as they search the prior art. Figure 10.2 shows the life cycle of technical information.

Books

Books are probably the most familiar scholarly information format for young engineers to use after years of textbook-based learning. Technical books typically are the culmination of extensive effort to summarize research and organize it into a coherent narrative, making them often the best source to consult when attempting to master the fundamentals of a topic or concept. Reference books, such as technical encyclopedias and handbooks, similarly summarize research findings from a variety of

sources, either core concepts or compilations of data. Encyclopedias typically only provide an overview of the topic, not at enough depth to gain competency, but enough so that the reader can get an idea of what a topic is about. Handbooks provide an easy way to access data from a variety of sources in one location. Books are increasingly available in electronic format, which allows for quick searching of the contents to find relevant passages.

When design teams begin reviewing books, engineering librarians can help identify subject headings that will produce effective catalog queries and help designers discover the prominent authors of the subject matter. A speedy gathering of materials is vital, so designers should quickly review features such as the table of contents and the indices of books to see if the book actually includes information directly related to the design project task. Whenever designers discover relevant books unavailable in either electronic or paper format, library staff can readily explain the procedures for accessing

Synthesize Possibilities

materials stored in repositories or shared collections, or which can be borrowed from other libraries.

Journals and Proceedings

Journal articles and conference proceedings should be accessed when looking for more current research results because they are the primary way that scientists and engineers formally communicate with each other about their latest discoveries and inventions. Therefore, browsing or searching the recent literature can inform designers of the state of the art of a particular field. Scholarly journal articles and conference papers can quickly be discovered using appropriate library indices and databases. Most libraries now offer tools that search multiple databases at the same time, and designers should leverage the value of these resources, while remembering that many advanced search functions are only available in a database's native interface. Students can optimize the speed of gathering appropriate articles by reading through abstracts, rather than the entire article, to determine relevance. Careful reading can then wait until after the gathering process is completed.

A type of scholarly article, commonly referred to as a review article, can be invaluable for designers during the search process because review articles identify the most prolific scholars and prevalent research trends related to any given technical topic, summarizing the state of the art at the time the article was written. Indeed, some journals only publish review articles. In addition to aiding designers in gaining a strong awareness of relevant research issues, review articles include bibliographies that can be mined to identify useful papers. Engineering librarians can help designers quickly determine the most relevant conferences that discuss topics related to their design project task.

Patents

Patents (see Chapter 5) are rich sources of information about engineered objects. In exchange for disclosing the form and function, and often the method of production, of an invention, the patent allows the inventor the exclusive right to commercialize the product for a period of time. Much of the patent literature never appears in journals or other formal literature, so neglecting the patent literature will leave a big hole in the design team's literature review.

Patents are legal documents, which means they can be challenging to read and to locate. Inventors don't necessarily want their patents to be found by competitors, so they use alternative language structures to describe their inventions (see Chapter 5). Consequently, a thorough patent search needs to include classification searching, as that provides the only uniform structure for characterizing inventions. A patent might be titled "Two-wheel human-powered transportation device" to obfuscate its true intentions, but it will be classified by the U.S. Patent and Trademark Office not only as a bicycle but, for example, by whether it has a side carrier, the arrangement of its wheels and steering fork, and whether it is collapsible or foldable. While commercial sites, such as Google Patents, provide quick and easy searches of the patent literature, and they can be good places to start to see what kinds of inventions are available, a comprehensive search can only be done using a structured database, such as the freely available U.S. Patent and Trademark Office's database (http://www.uspto.gov), and Espacenet (http://worldwide.espacenet.com), which indexes patents from several countries.

Engineering librarians can play an invaluable role in helping students get started efficiently with their patent research by selecting the best database to search, by guiding students in selecting appropriate classifications, and by selecting appropriate assignees and inventors within queries to help focus searching. Identifying the assignees of patents is extremely important because designers can then seek out relevant product information from other company information sources. Patents are a crucial type of technical literature to search for design projects because most, if not all, patents include state of the art summaries (i.e., mini–literature reviews). Designers can quickly gather abstracts and read the claims, which explain what exactly the patent is protecting, to select patents for further review.

Standards

Technical standards are probably the least familiar type of technical literature for capstone design teams, and some students may never have read a standard prior to their first major design project. The value of standards for design projects cannot be overstated because these information sources entail best practices for products and processes, essentially the collective wisdom of a variety of experts who have thought deeply about a topic over an extended period of time. (See Chapter 9 for more information about standards.) Standards should not limit designers, but rather provide structure for the set of requirements and test methods their project may need to fulfill, related to whatever types of materials, systems, components, or processes are pertinent to their project task. Standards can be readily accessed via library catalogs and databases, and they can be quickly selected by students after they read the scope of the standards, similar to an abstract, at the beginning of the document. Engineering librarians can be helpful at the start of the query process by identifying relevant types of standards for specific design projects, and, since standards

are produced by many different organizations, librarians will know the best way to access a particular standard. Designers should also ask the key stakeholders for guidance because they will probably have a strong awareness of their industry compliance issues.

Product/Trade Literature

Popular literature provides vivid, easily readable (and viewable) content for inspiration during the brainstorming phase of solution synthesis. It is easy to locate a large volume of popular and trade literature via a general Internet search. However, since this information is very informal and fluid, and often has as its primary purpose to sell a product (i.e., only stating what a product does well and not what its limitations are), students need to use their evaluation skills to determine what information is actually contained in a particular resource and how they can independently verify the veracity of that source. (See Chapter 11 for strategies.) In particular, students often locate what they think is the perfect part for their project by doing a quick Web search. However, they may only read the headline "most energy efficient fluorescent bulb on the market," without realizing that the advertisement is for a T1 style (three-foot-long) bulb, rather than a compact fluorescent that would be more appropriate for the personal reading lamp they are designing.

Students can be savvy about navigating trade literature by locating product spec sheets, manuals, and warranty details to see exactly how and how well a product works. Similarly, locating review sites, both consumer sites as well as industry magazines and blogs, will help students determine whether a product meets the specifications it alleges. Industry magazines and blogs can also highlight new technologies and popular products and can provide

Synthesize Possibilities

inspiration for looking at a design problem or for querying the formal literature in new ways.

TEAM PROCESSING OF PRIOR ART

Finding an initial quantity of diverse and relevant scholarly literature is one matter, but design teams will also need to read and understand the information as they conduct a thorough search of prior art. An effective practice involves design team meetings in which designers divide up the reading material and report on what they have read. Each team member then reports on the items he or she read with summaries that are three minutes or less in length. Whenever possible, the source of information should be displayed with a projector as designers deliver their summaries. For the sake of efficiency, all literature summaries should be delivered with the same key elements. A simple and effective approach involves answering a list of basic questions such as the following:

- What did you read?
- Who created the information?
- Why do you think it is credible?
- Why is it valuable for the project?
- How can you use the information in the design process?
- Should your fellow team members read it?
- Does it raise important questions to ask your advisor?
- Does it identify a need for more reading materials?

This can even be carried out as a small-group activity within the classroom, with instructors and librarians helping facilitate discussions among team members.

As decisive documents of value are identified, additional time can be provided for the team to observe the related figures as a group. Compiling the literature in a shared citation manager (see Chapter 6) will help the team keep all information organized and accessible.

This approach is particularly effective with patents because the detailed figures required within patents to define the processes and features of inventions provide an ideal way for designers to visualize prior solutions. In addition, once valuable information is identified, designers can take advantage of bibliographies from those sources to identify even more sources. Design teams ought to engage in follow-up meetings with engineering librarians, who can then offer practical recommendations about how to expand their searching efforts.

SUMMARY

The interconnectivity of the technical literature will become apparent to design teams as they engage in the search process. For example, a design team might discover a relevant manufacturing company they did not know about as they examine a patent in which the company is identified as the patent's assignee. In addition to searching for all of the valuable patent information related to the company, the design team can then access information about the company's technical product information via the Web. Likewise, the name of an executive engineer identified in a press release may serve as the basis of a query to find an associated patent. A press release might also indicate an important compliance issue for a specific standard the design team had not yet considered for their search. Marketing brochures might indicate technical specifications, warranty details, and product testing results that designers might not discover through reading patents and standards. Online demonstration videos of products and processes might indicate details previously unknown to them. When design teams engage in this type of detective work, they develop a considerable expertise for making strong decisions further ahead in the design cycle process.

SELECTED EXERCISE

Exercise 10.1

A major league baseball player wants a maple baseball bat with the widest sweet spot, the lightest weight, and the strongest durability possible that is also legal for professional use. Have students brainstorm what kinds of scholarly and popular literature can be used to search the prior art for this topic. Have them discuss the possible information sources that could inform their knowledge and divide up the different literature types among the various team members. Each team member then spends 30 minutes searching for information in the source assigned to him or her. The team members read the materials they found independently and meet at a later time to report to each other, in 3 minutes or less, on what they learned. Have students determine which types of literature were the easiest and hardest to find, and which sources, if any, surprised them. Have them identify which types of information the team would look for if they were to continue their search.

REFERENCES

Dahl, D. W., & Moreau, P. (2002). The influence and value of analogical thinking during new product ideation. *Journal of Marketing Research*, *39*(1), 47–60. http://dx.doi.org/10.1509/jmkr.39.1.47.18930

De Bono, E. (2009). *Lateral thinking: A textbook of creativity*. London: Penguin Books.

Frog Design. (2013). *Collective action toolkit*. Retrieved from http://www.frogdesign.com/collective-action-toolkit.

Golembewski, M., & Selby, M. (2010). Ideation decks: A card-based design ideation tool. In *Proceedings of the 8th ACM Conference on Designing Interactive Systems* (pp. 89–92). New York: Association for Computing Machinery. http://dx.doi.org/10.1145/1858171.1858189

Kelley, T., & Littman, J. (2001). *The art of innovation: Lessons in creativity from IDEO, America's leading design firm*. New York: Currency/Doubleday.

Kolodner, J. L. (Ed.). (1993). *Case-based learning*. Dordrecht, Netherlands: Kluwer Academic Publishers.

Morgan, M. (1993). *Creating workforce innovation: Turning individual creativity into organizational innovation*. Chatswood, NSW, Australia: Business & Professional Publishers.

Strouse, R., & Pollock, D. (2009). *Understanding scientists' and engineers' information use habits, preferences, and satisfaction*. Burlingame, CA: Outsell, Inc.

Wang, H.-C., Cosley, D., & Fussell, S. R. (2010). IdeaExpander: Supporting group brainstorming with conversationally triggered visual thinking stimuli. In *Proceedings of CSCW 2010*. New York: ACM Press.

Wodehouse, A., & Ion, W. (2012). Information use in conceptual design: Existing taxonomies and new approaches. *International Journal of Design*, *4*(3), 53–65.

Synthesize Possibilities

CHAPTER 11

MAKE DEPENDABLE DECISIONS

Using Information Wisely

Jeremy Garritano, Purdue University

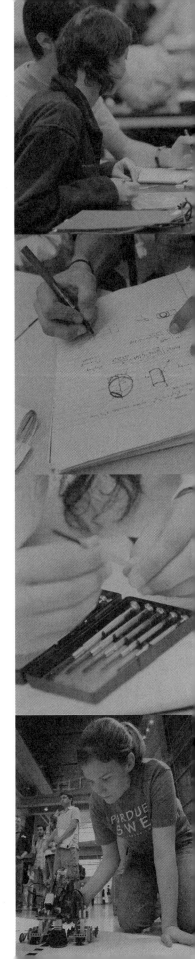

Learning Objectives

So that you can guide students to think critically about information they locate to support a design project, upon reading this chapter you should be able to

- Outline the major challenges student design teams have in determining the quality of information from various sources

- List and describe the importance and significance of six criteria for determining the trustworthiness of information

- Explain the application of three techniques for evaluating the quality of potential or proposed solutions in order to make dependable decisions

INTRODUCTION

Having synthesized knowledge of the specific needs of the stakeholders (Chapter 7), the context of the design task (Chapter 8), professional requirements and best practices for performance (Chapter 9), and the universe of previously developed solutions (Chapter 10), student teams will then systematically choose the solution that best fits their situation. This is an important step in the design process because

- designers can drive further efficiency or economy in implementation by comparing their ideas and solutions to those of others;
- designers will spend less time in testing or deployment since they will have eliminated less promising solutions and false leads early on in the process;
- aligning solutions with stakeholder needs will improve stakeholder satisfaction and acceptance of the final design solution.

The selection of potential solutions relies on evaluating the solutions on both nontechnical and technical bases. A number of evaluation and comparison activities, in order of increasing complexity, are discussed in this chapter.

COMMON CHALLENGES FOR STUDENTS

Students are aware that there are differences in information found on a freely available website versus a library database. A study by Head and Eisenberg (2010) confirms that students scrutinize public websites (seven or more evaluation standards used) more than library materials (four or fewer standards used). However, for students, the justification of the quality of an information resource can still be very shallow, even simply, "I know good information when I see it." While various criteria for examining the trustworthiness of a source might seem obvious (e.g., who wrote it, what are their credentials, how old is the information), students may not slow down long enough to consider each criterion. A recent study indicates that undergraduate students do "not necessarily apply the selection criteria that they claimed to be important" (Kim & Sin, 2011, p. 184) when evaluating information resources. Also, in the digital age it can sometimes be difficult to identify all of the criteria for a particular source.

Using databases that offer easily identifiable fields such as the author, author's organization, and date of publication are a great help compared to searching the open Web through a search engine. When comparing potential solutions, students may also have difficulty in extracting the technical information necessary to compare the solutions on the same level. Students are not experts in the field, and reading technical literature can be daunting. Additionally, not all of the needed information is usually found in one source, so students often need to piece together information from multiple sources in order to conduct a thorough analysis. There will also be gaps in knowledge, and students become frustrated when they find information related to one solution—say, monetary cost or environmental impact—but cannot find it for another. Finally, while not the same as a gap in knowledge, the ability to distinguish latent information versus explicit data described in a solution can also present a challenge for students. Not all conditions can be investigated during an experiment, so even if a solution or piece of equipment seems viable given favorable results in an article or report, it may not be able to withstand the particu-

Select Solution

lar environmental conditions of the new application—for example, if the team is designing for an environment that is extremely cold or exposed to high levels of moisture. When evaluating potential solutions, it is also important to be able to read between the lines and see what assumptions might have been made, even if unintentional. As an example, materials tested outdoors in the Southern United States might rarely see below-freezing temperatures and could be problematic for installation in the Northeastern United States.

EVALUATING THE TRUSTWORTHINESS OF INFORMATION

Potential solutions gathered from various sources often vary widely in their degree of quality. Any information used in the process of evaluating potential design solutions must be vetted for its trustworthiness and authority. Six basic criteria—authority, accuracy, objectivity, currency, scope/depth/breadth, and intended audience/level of information—used to do this are discussed below. These criteria have been adapted and expanded from a list of five criteria for evaluation of Internet resources suggested by Metzger (2007).

Authority

Students must consider the author/creator of the source, including credentials, qualifications, how closely they are associated with the original research, and whether they have been sponsored or endorsed by an institution or organization.

In finding research articles related to current technologies for distillation columns, how accepting of the claims of column efficiency should a student be if the author were a process engineer working at a petroleum company? A sales person working at a company that manufactures the columns being described? A chemical engineering professor at a university that has a lengthy history of publishing on column efficiencies?

A student finds a potential solution for increasing solar cell efficiency from a trade magazine. Is the author of the article a journalist reporting about the solution or is the author the originator of the solution? The student should follow the path back to the original research to read about it firsthand.

Accuracy

Students must consider whether the conclusions are appropriate and consistent given the wider body of knowledge and whether the claims made are supported by the evidence provided.

For many research publications, students should pay attention to sections such as the introduction, literature review, background, and conclusion, to see how authors are characterizing their work compared to that previously reported. Claims of breakthroughs or results inconsistent with past research may need to be verified by additional sources that confirm the initial claims. Bibliographies or works cited lists can be consulted for additional verification.

Objectivity (of Both the Author/Creator and the Publisher)

Students must consider whether the author/creator/publisher has a mission/agenda/bias that would raise doubts as to the credibility of the information and determine whether there any conflicts of interest such as funding sources, sponsoring organizations, or membership in special interest groups.

In researching existing technologies and safety issues related to hydraulic fracturing, a student finds reports from the EPA (Environmental Protection Agency), Chevron, and a website called The True Cost of Chevron. How would knowing that the EPA is a government organization charged with investigating and reporting on environmental issues, that Chevron is a company that conducts hydraulic fracturing, and that the final website is supported by a variety of nonprofit organizations protesting hydraulic fracturing impact the student's view of the objectivity of each report? How might the student reconcile contradictory information?

Currency

Students must consider not only the date when the information was published but also the date when the data was actually collected. Would an older solution continue to meet standards, laws, and regulations enacted since its publication? Should older solutions be reexamined in the context that these solutions may have been initially overlooked or are now considered viable given current technologies or social/economic/political trends?

Review articles, while useful, may cover a wide range of research published over a decade or more. When referencing tables or figures that are published in these articles, students must be careful to note when the actual data was published if the author is reprinting or collecting previously published data.

Students require guidance on what is considered current in their discipline. Knowing how quickly the electronics field makes advances, would a report on semiconductors that is 5 years old be considered current? What if

the report were 10 years old? What about in other rapidly advancing fields such as nanotechnology or biotechnology?

Scope/Depth/Breadth

Students must consider how specific the solution is compared to the desired application and under what variety of conditions the solution has been tested or implemented in order to extrapolate its applicability.

A student may find a report of new jet fighter wing designs in a conference proceeding. The student should be careful in extrapolating the solution's appropriateness, as the purpose of some conference presentations is to present preliminary results to the engineering community that may not be fully tested, especially across a wider range of variables (such as particular speeds, temperatures, or altitudes) that may be important to the student's artifact.

Intended Audience/Level of Information

Students must consider the intended audience of the information source, which may be written for the general public, an organization of professionals, or government officials. How do different audiences affect the presentation of the solution?

A solution a student may find described in a popular science and technology publication such as *Scientific American* may be oversimplified since its audience is meant to be the general public. The description may be incomplete, especially regarding specific details that would be required to truly compare the solution against others gathered.

Having students search in quality databases, such as those provided by libraries through institutional subscriptions, can often reduce the amount of time students must spend evaluat-

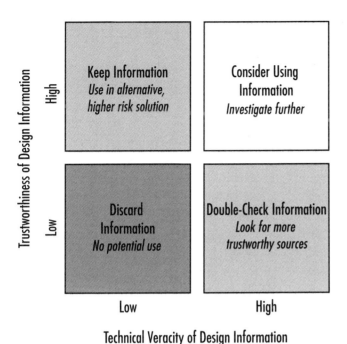

FIGURE 11.1 Design information decision grid.

ing potential solutions. Results from searches on the Internet through general search engines, on the other hand, deserve enhanced scrutiny using the previously mentioned evaluation criteria. In situations where students may not have as much technical background to truly evaluate potential solutions, evaluating some of these nontechnical aspects can be just as useful in narrowing down a lengthy list of results. (See Hjørland [2012] for a concise summary of 12 ways in which information sources can be evaluated.)

ASSESSING THE CONTEXTUAL APPLICABILITY OF DESIGN INFORMATION

To be useful, information must have technical relevance in the particular design context. The types of questions that get at the technical relevance of information include the following:

- Is this the appropriate technical information for the design decision at hand?
- Has this technology (concept, material, component, etc.) been used successfully in a comparable context? Or is this a new, untested technology?
- Does this technology address the needs of the client and other stakeholders?
- Are there negative social or environmental aspects to this technology?
- What are the life cycle costs associated with this technology or design solution?

Broadly stated, a student can plot the potential value of a piece of design information along a continuum of how trustworthy it is and how relevant it is to the particular design problem. The essential design decision about whether or not to use particular information is depicted in Figure 11.1.

The particular course of action students should take depends upon in which of the four quadrants a particular piece of design information is located. For example, if the design idea or technology found is based on untrustworthy information *and* is deemed to have low relevance to the design task at hand, it can be deemed not viable and thus discarded from further consideration. Conversely, information from trustworthy sources that offer highly relevant solutions deserves further consideration and additional information might need to be sought.

If the idea or technology is highly relevant and shows high technical potential, but it comes from an untrustworthy source (let's say, a blog), then the student should proceed cautiously and definitely seek confirmation of the technical potential from additional information sources that are trustworthy. For example, the blog post might have mentioned published research, or the author of the blog post might be a reputable researcher or a designer with a proven track record. In this case the student could track down the original research using an author search in a library database. Conversely, if the information comes from a trustworthy source but is not particularly relevant to the context, then the student should keep the information for further consideration, possibly for use in an unconventional approach that, while it is unproven (and thus is riskier), might provide a more innovative, game changing design solution. An example of this might be a student investigating recycling efforts on college campuses. A peer institution might have a successful recycling program but not have a print student newspaper. So, unlike the student's campus, the peer institution does not need to recycle newsprint. While coming from a high-quality source, the peer institution's solution does not handle all situations being investigated by the student. The peer's program may be investigated for particular aspects of the solution, but as an overall program it is not the best match.

Potential solutions gathered from various sources often vary widely in their degree of overall quality—defined as the combination of trustworthiness of the information and the applicability. Any information used in the process of evaluating potential design solutions must be well documented and recorded for appropriate comparisons to be made. What follows are three methods for comparing the quality of various solutions in order to narrow down the solutions to be considered. Each method is more sophisticated than the next and therefore would require students to have correspondingly more accurate, detailed, and trustworthy information about each potential solution.

Method 1: Pro/Con Evaluation

In Method 1, potential solutions are listed in a table with separate columns related to the pros and cons of each solution (Pahl & Beitz, 1996). An example is the rehabilitation or replacement of an aging bridge across a river. If there are actually two bridges, one for traffic in each direction, there are a variety of ways the bridges can be rehabilitated or replaced (see Table 11.1).

Only minimal and not necessarily complete information is needed for each possible solution. This method provides a very simple way to compare potential solutions on a rough scale and can reveal some general trends of the strengths and weaknesses of alternatives, but it does not offer a more data-driven or objective analysis.

Method 2: Pugh Analysis

Method 2, a Pugh Analysis (Pugh, 1991), can take information in a format similar to that of Method 1 but will compare each potential so-

TABLE 11.1   *Pros and Cons Evaluation of Rehabilitating or Replacing an Existing Bridge*

Design/Solution	Pros	Cons
Rehabilitate existing bridge	Cheapest option Least disturbance to local geography	Lowest estimated service life Existing bridge would need to be thoroughly analyzed before repair Traffic diverted to other bridge during rehabilitation
Remove existing bridge; rebuild on same alignment	Longest estimated service life	Traffic diverted to other bridge during construction
Remove existing bridge; build to another alignment	Longest estimated service life No traffic restrictions during construction	Highest cost option Greatest disturbance to local geography

lution to either the current situation or a proposed solution the student wants to compare all other solutions against. More specific information is needed about each solution, as the student will then rate each criterion of a new solution against the existing solution or an initial proposed solution—in this case, a "+" for better than the baseline solution (existing or initial proposal), a "–" for worse than the baseline solution, or an "s" for same as the baseline solution. These are then summed to give a final score, and the results can then be reflected upon. In the case of the bridge rehabilitation, if the solutions are compared against simply rehabilitating the existing bridge, a Pugh Analysis might look like the analysis shown in Table 11.2.

To create this table the student would need to know detailed information on costs and service life, for example, in order to determine whether the solution criteria were better or worse than the proposed solution. Looking at the summations gives a more objective idea of how the alternative solutions compare to the proposed solution over the pro/con analysis.

Method 3: Weighted Decision Making

Method 3 takes an analysis similar to the Pugh Analysis but adds the dimension of weighting the criteria to further align the needs of the stakeholders with the proposed solutions (Cross, 2008; Pahl & Beitz, 1996). This is especially helpful if there are no clear winners among a Pugh Analysis. (For example, in Table 11.2, there are differences between the two proposed solutions, but it could be argued that there is not a clear alternative that is better than the other.) There are eight steps to constructing a weighted decision matrix:

1. List criteria (based on stakeholder needs).
2. Weight these criteria.
3. Determine metrics: What will be measured to determine if each criterion has been met?
4. Determine targets: Is there an optimal value for some of the metrics? What is the optimal value? (For some metrics, there will not be a target value.)
5. Determine relationships between criteria/needs and metrics: There might be one metric

TABLE 11.2 *Pugh Analysis of Rehabilitating or Replacing Existing Bridge*

Criterion	Proposed Solution: Repair Existing Bridge	Alternative 1: Rebuild on Same Alignment	Alternative 2: Rebuild to Another Alignment
Approaches realigned?	No	s	–
Estimated service life	10 years	+	+
Traffic restrictions during construction	1 lane, northbound and southbound	s	+
Cost estimate	$8 M	–	–
Sum (+)		1	2
Sum (–)		1	2
Sum		0	0

NOTE: "+" means the criterion is better than the proposed solution; "–" means criterion is worse than the proposed solution; "s" means the criterion is the same as the proposed solution. These are then summed: "+" = 1, "–" = −1, and "s" = 0.

Select Solution

for each criteria, one metric that addresses multiple criteria, or several metrics that measure different dimensions of a single criterion. Use an "x" to denote that a metric is related to a particular criterion. If there are no metrics related to a particular criterion, add an additional metric.

6. Give scores to the alternatives based on actual data, whether gathered from existing research or determined by experiment/prototype.

7. Calculate the weighted total for each alternative: First calculate the weighted score for each criterion for each alternative, then sum the weighted total for each alternative.

8. Reflect on the results: Do they make sense?

This approach offers the potential for objectivity, if the weights are determined without any particular solution in mind, ideally using information gathered from stakeholders to determine the criteria and weights (see Chapters 7 and 8). In the bridge example, perhaps it

is determined that due to other construction projects going on within the city, it is necessary to minimize traffic disruptions. Therefore the criterion "traffic restrictions during construction" (see Tables 11.1 and 11.2) will carry more weight than others. Additionally, costs are often a factor, so that criterion may also carry a greater weight. If the eight steps are followed as described, a weighted decision matrix (see Figure 11.2) will result. In the bridge example, as shown in Figure 11.2, because of the various weights given to the criteria, the solution "rebuild to another alignment" ends up with the highest score. Students would need to reflect then on what the scores really mean and if it makes sense that this appears to be the best solution to pursue. If it does not, then the weights might be reviewed and/or additional information and metrics could be added to the analysis if gaps are identified. Students must be cautious not to make modifications in order to raise the score of the solution that is

Criteria	Weight/importance	Metric				Unweighted, scale of 1–5 (5 being most aligned with design criteria)			Weighted scores		
		Yes/no	Service life	Yes/no	Cost	Repair existing bridge	Rebuild on same alignment	Rebuild to another alignment	Repair existing bridge	Rebuild on same alignment	Rebuild to another alignment
Approaches realigned?	10	×				5	5	1	50	50	10
Estimate service life	10		×			2	5	5	20	50	50
Traffic restrictions	15			×		1	1	5	15	15	75
Cost estimate	15				×	5	4	4	75	60	60
Engineering targets →		No	>25	No	$8				**Totals**		
Units		Non-dim	Years	Non-dim	$M				160	175	195

FIGURE 11.2   Weighted decision matrix for rehabilitating or replacing existing bridge.

Select Solution

simply preferred by either the designer or the stakeholders. The purpose of this matrix is to maintain as much objectivity as possible.

EVALUATING WHEN THERE ARE GAPS IN KNOWLEDGE

Given the three methods discussed in the previous section, any process in which data is placed into a carefully ordered grid or table might imply that a student will then be able to quickly read the table and decide what the best solution is, even if no weighted decision making is involved. In reality, an analysis often contains gaps, and these gaps are where students can struggle. One of the main questions for students to answer is, *Do I have sufficient information that I trust in order to make to make a design decision that I can stake my reputation on?* For example, in comparing solutions, one of the criteria might relate to comparing the environmental impact of the potential solutions. From the data gathered, it might be extremely difficult to know this information about every solution, since some solutions might still be in development, have test results that are confidential, or not even be fully implemented, especially in the context one is considering. To assist students in these gray areas, it is important to emphasize using existing knowledge and stakeholder needs to decide whether

- the particular gap in knowledge must be filled in order to continue. This might involve further searching for evidence or even calling up the particular people or company responsible for the solution in order to gain the necessary information.

- assumptions can be made. Knowing how similar solutions behave, can an assumption be made regarding how one particular solution will behave compared to another known solution?
- the gap in knowledge can be ignored. In the end, is the particular gap deemed not as important, or would it not factor into the desirability of the solution, so that the information is not necessary?
- stakeholders must be consulted. Is there enough uncertainty in the gap in knowledge that the stakeholders must review the importance or weight of the particular criterion in question?

Kirkwood and Parker-Gibson (2013) have detailed two comprehensive case studies for researching engineering information related to ecologically friendly plastics and biofuels, including evaluating information resources as a search progresses.

ACKNOWLEDGING SOURCES OF IDEAS

Once a set of potential solutions has been identified for further exploration, it is also important to acknowledge the sources of those ideas throughout the design process. Stakeholders should be informed of sources in order to provide feedback or reveal any additional knowledge or conflicts of interest given the selected potential solutions. If the solution is to be commercialized or pursuit of intellectual property protections are desired, it is important to document the prior art in order to determine what is original and what is already known. Intellectual property concerns may also prove to be obstacles in implementing or modifying exist-

Select Solution

ing solutions if particular solutions are still under protection and may require licensing from the patent assignees. Also, when evaluating the quality of proposed solutions, if analysis of criteria is undertaken, such as through a Pugh Analysis or weighted design matrix, it will be necessary to document the source from which each criterion was derived. For the bridge example, information on the life of a new bridge may have come from a source different than the one that detailed the costs of the new bridge. Information on the life span of the rehabilitated and new bridges may have come from different sources that used different methods for calculating anticipated lifetimes. In these cases, it would be important to annotate or cite the source of each criterion in case the original source would need to be referenced again. Particular tools that can manage citations have been previously discussed in Chapter 6. In the case of acknowledgment, the emphasis should not be placed on mastering any one particular citation style. Instead, the emphasis should focus on being consistent in the use of citations and in the way they are presented, regardless of the style used.

SUMMARY

In this chapter we considered how information such as stakeholder needs, the context of the design task, and prior published work addressing similar problems can be used as inputs in order to select the most promising potential solutions for further consideration, as well as to compare these solutions to a current or proposed solution. We reviewed a list of criteria for evaluating the trustworthiness of a source as well as several techniques for comparing solutions based on their technical details. Once students identify the most appropriate approach, they can start

to work on embodying their solution—that is, determining how they will actually implement their solution. This will involve gathering more detail-oriented information, such as selecting materials and components that will meet the design requirements, as discussed in the following chapter.

SELECTED EXERCISES

Exercise 11.1

As pre-work for a class, have students research a particular topic, such as efficiency of wind turbines or biodegradability of particular polymers, and collect what they feel are five highly relevant information sources. Have students annotate the resources using the six criteria discussed (authority, accuracy, objectivity, currency, scope/depth/breadth, and intended audience/level of information) to justify their relevancy. In class, in small groups have students discuss with each other their top source, their rationale for picking this source, and what aspect of the quality of their source they are most uncertain about.

Exercise 11.2

For a particular design problem, have students independently research potential solutions creating their own pros and cons list. Then in class, within design groups, have the students analyze the potential solutions, creating a Pugh Analysis or weighted decision matrix (depending on the complexity of the assignment and level of detail you require) to turn in by the end of class. Students will need to work together to agree on what solutions are better than the current model, as well as potentially create different weights, measures, and targets based

Select Solution

on existing knowledge, including information gathered from clients.

ACKNOWLEDGMENTS

The author acknowledges Monica Cardella, Purdue University, for the use of her instructions for constructing the weighted decision matrix. Data for the bridge repair/rehabilitation analyses was adapted from a report made by Parsons for the Indiana Department of Transportation, US 52-Wabash River Bridge Project, Des. No. 0400774, http://www.jconline.com/assets/PDF/BY1656611019.PDF, accessed July 3, 2013.

REFERENCES

Cross, N. (2008). *Engineering design methods: Strategies for product design* (4th ed.). Chichester, West Sussex, England: John Wiley & Sons Ltd.

Head, A. J., & Eisenberg, M. B. (2010). *Truth be told: How college students evaluate and use information in the digital age*. Project Information Literacy Progress Report, The Information School, University of Washington. Retrieved from http://projectinfolit.org/pdfs/PIL_Fall2010_Survey_FullReport1.pdf

Hjørland, B. (2012). Methods for evaluating information sources: An annotated catalogue. *Journal of Information Science, 38*(3), 258–268. http://dx.doi.org/10.1177/0165551512439178

Kim, K.-S., & Sin, S.-C. J. (2011). Selecting quality sources: Bridging the gap between the perception and use of information sources. *Journal of Information Science, 37*(2), 178–188. http://dx.doi.org/10.1177/0165551511400958

Kirkwood, P. E., & Parker-Gibson, N. T. (2013). Informing chemical engineering decisions with data, research, and government resources. In R. Beitle Jr. (Ed.), *Synthesis lectures on chemical engineering and biochemical engineering*. San Rafael, CA: Morgan & Claypool Publishers. http://www.morganclaypool.com/doi/abs/10.2200/S00482ED1V01Y201302CHE001

Metzger, M. J. (2007). Making sense of credibility on the Web: Models for evaluating online information and recommendations for future research. *Journal of the American Society for Information Science and Technology, 58*(13), 2078–2091. http://dx.doi.org/10.1002/asi.20672

Pahl, G., & Beitz, W. (1996). *Engineering design: A systematic approach* (K. Wallace, L. Blessing, & F. Bauert, Trans., 2nd ed.). K. Wallace (Ed.). London: Springer-Verlag.

Pugh, S. (1991). *Total design: Integrated methods for successful product engineering*. Workingham, England: Addison-Wesley.

Select Solution

CHAPTER **12**

MAKE IT REAL

Finding the Most Suitable Materials and Components

Jay Bhatt, Drexel University

Michael Magee, Drexel University

Joseph Mullin, Drexel University

Learning Objectives

So that you can advise your student design teams on what information sources are available to help them turn their design concepts into reality, upon reading this chapter you should be able to

- Describe and illustrate the major challenges student design teams face in finding and then deciding between the multitudinous options available when they have to select materials and components

- List the major factors that should be considered when selecting a material for fabrication or commercial off-the-shelf components or systems

- Demonstrate effective and efficient strategies for selecting the most appropriate materials to use in fabricating a new product

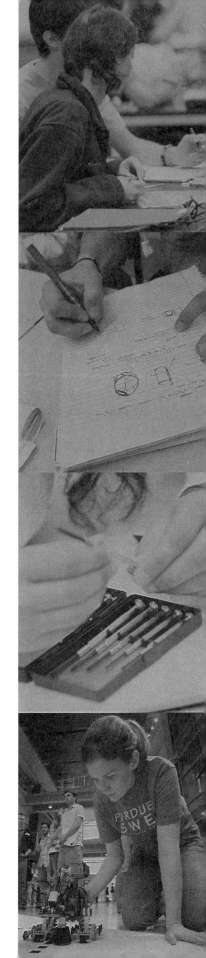

INTRODUCTION

The previous stages of the design process have helped determine what the students' artifact needs to do, how well it needs to do it, and possible ways to accomplish this. Once a preferred concept to solve the design problem has been selected, the details of how to actually build the artifact must be determined and embodied in the final artifact.

Selecting the most appropriate and cost-effective materials and components is critical to the success of a design project (Ashby, 2011a). Without careful materials selection, the resulting artifact may be suboptimal in terms of performance, ease of manufacture, fabrication, or cost (Jahan & Edwards, 2013). A disciplined and methodical investigation of alternative ways to realize the concept is necessary in order to create competitive, cost-efficient design solutions. Embodying a design concept includes considerations of both the materials used and how these materials will be shaped or otherwise transformed into the manufactured artifact. For example, if a particular type of metal is too brittle to be extruded in a manufacturing process, even if it has the appropriate mechanical properties, it may not be appropriate for use in the final project.

This chapter describes a general process for materials selection and a discussion of strategies and resources for locating materials. When searching for information, students need to determine the most important sources for finding material properties and assess the reliability of those sources. In many cases embodiment of a concept is achieved in part through the selection of existing commercial off-the-shelf components (COTS); therefore, consideration is also given to finding information on the performance and other specifications of COTS.

COMMON CHALLENGES FOR STUDENTS

Students can be overwhelmed by the vast number and variety of materials available to them. Whether it is the hundreds of different kinds of steel available on the market, or the multitude of chipsets produced by dozens of manufacturers, students struggle to locate materials or components relevant to their need. They often take the first material that looks reasonable, perhaps the first item that shows up on an Internet search, rather than trying to systematically find the best material for the job.

Materials specifications and data sheets often contain large amounts of difficult to understand technical detail, and consequently, students have considerable difficulty in sorting through and interpreting the voluminous data they do find, or knowing how to distill or translate this into usable design information. This is made all the more difficult if the student does not have a thorough grasp of fundamental concepts in material properties and how these relate to material behavior (e.g., Young's modulus, conductivity, flexibility, or rigidity). An artifact being designed typically has multiple components. The materials for each component must be carefully selected so that the assembly performs properly in the final product.

For example, a swimming pool diving board has limitations on size, load capacity, and deflection when in use. Further, it must resist the dynamic loads that a diver applies to it in performing a series of dives. Its ability to store strain energy like a spring is a critical parameter. Indeed, this is perhaps the most important function that a diver values in the board's design, as it translates into the ability to spring high into the air when beginning the act of diving. The diving board must provide this rebound energy

REALITY CHECK 12.1

Role of Materials in Successful Engineering Design

Materials play a critical role in successful engineering design. Proper material selection can sometimes decide whether or not a system is designed so that it is safe to the users and the public. In December 2012 a shark tank in a Shanghai shopping center collapsed just two years after it was constructed, injuring 16 people and killing the sharks and dozens of other sea animals it housed. Investigators concluded that two years of UV light exposure from the sun and thermal cycling from the outdoor climate had caused the 10-inch-thick acrylic glass panel to become brittle enough to crack (Ho, 2012).

with minimal deformation and without excessive vibration. So it must be a finely tuned cantilever beam, light and stiff on the one hand, yet able to quickly damp out vibration after the dive is complete (Chopra, 2012).

MATERIAL SELECTION STRATEGY

In order for students to be able to search effectively, they first need to know what it is they are looking for. Often they haven't sufficiently determined the precise problem they are trying to solve (e.g., the performance requirements their component needs to meet), and without clearly understanding the problem, students have difficulty recognizing a viable solution.

The following question-based strategy for material selection and COTS component selection can be used by students to overcome many of the difficulties they often experience when embodying their design concepts.

1. What performance is required from the component?

2. What are the environmental factors across the life cycle of the artifact?
3. Are there commercially available components or products that will do the task?
4. What relevant information is needed to be able to select a suitable material?
5. What materials are potential candidates for this application?
6. Are there newer materials or technologies that might offer innovative design solutions?
7. What materials selections charts or software are available?
8. What form and size do the materials come in?
9. How will the materials be processed or shaped in order to make the component?
10. Are there other constraints related to the materials that must be satisfied?

Various classes of materials are available, and each class contains many different types of materials (see Table 12.1).

Through the use of a variety of materials based on properties, applications, cost, and

REALITY CHECK 12.2

Designing a Green Roof

A lightweight vegetated roof research team was challenged with finding a material for their substrate medium. In addition to common properties desirable in similar applications, environmental impacts such as resource extraction, total embodied energy in production and distribution, and disposal were most important to them. They first made a list of possible material choices based on bulk density, durability, and absorptivity, then each material was put through a life cycle analysis, which revealed information about sourcing and the process required for manufacturing. For example, EPS (Expanded Polystyrene) had excellent properties that would work well for their system; however, due to its large embodied energy and the fact that it is not biodegradable, it had to be eliminated as a candidate.

TABLE 12.1 *Classes and Examples of Materials*

Class	Material
Metals and alloys	Iron, steel, copper and alloys, aluminum and alloys, nickel and alloys
Polymers	Polyethylene (PE), polymethylmethacrylate (acrylic and PMMA), nylon or polyamide (PA), polystyrene (PS), polylactic acid (PLA)
Ceramics and glasses	Alumina (Al_2O_3, emery, sapphire), magnesia (MgO), silica (SiO_2) glasses and silicates, silicon carbide (SiC)
Composites	Fiberglass (GFRP), carbon-fiber reinforced polymers (CFRP), filled polymers
Natural materials	Wood, leather, cotton/wool/silk, bone, rock/stone/chalk

Data from Ashby & Jones, 2012.

other factors, the most appropriate materials can be selected in order to design and develop the final product.

ENVIRONMENTAL CONSIDERATIONS

In the diving board example from the previous section, because the board must operate in a very moist environment, if wood is used in this application it has to be resistant to damage when constantly wet or exposed to wet/dry cycles. This often requires sealants on the wood to keep it dry. It also requires that the hardware used to mount the board on a diving platform must resist any form of corrosion. Galvanized steel was the normal standard in wet environments. Similar design constraints were set on boats made of wood. Steel fasteners were usu-

ally galvanized (coated with zinc) to resist corrosion (Dowling, 2007).

In contemporary diving board design, wood has been replaced by fiberglass. Glass fibers in epoxy are much lighter and stronger than wood and can be formed into the specific shapes most efficient in providing the desired performance characteristics. These new composite materials can be optimized as to strength, stiffness, ability to store more energy, and even improved damping characteristics. There is very little water penetration and therefore no need for sealants, although some are painted and coated with a gel coat of epoxy resin, giving them a very smooth and attractive appearance. Fiberglass, unlike carbon fiber reinforced resin, is not terribly expensive and is therefore broadly used in marine applications (Masuelli, 2013).

COMMERCIAL OFF-THE-SHELF (COTS) COMPONENTS

When selecting materials, it is also necessary to determine whether any COTS components should be used in the product design. While many engineering students think first of designing their own custom solution to a problem, down to the individual parts, custom designed components may be prohibitively expensive to produce in quantity with marginal increase in efficiency and performance of the final product.

The market provides access to a variety of available COTS. In the overall design process, these components can play an important role in the successful design project. According to Farr (2011), "A commercial off-the-shelf (COTS) component is an item bought from a third party supplier and integrated into a larger system" (p. 207). Some examples of COTS

Refine Solution

Commercial Off-the-Shelf Components

Companies such as Adafruit Industries (www.adafruit.com), SparkFun Electronics (www.sparkfun.com), and Maker Shed (www.makershed.com) sell low cost COTS software and electronic parts, such as the Italian microprocessor Arduino. These materials are extremely useful for low cost prototyping. There are extensive professional and hobbyist communities that provide an abundance of freely available information and open source scripts that can perform various prototyping functions. Students at the Drexel Smart House in Philadelphia use the Arduino platform paired with various flow meters, sensors, and servos to control an indoor farming prototype. This allows them the ability to quickly change microprocessing controls, which gives them the flexibility to efficiently experiment with many different program settings of the automated system toward finding the most optimal system design at a low cost.

components include computer software, hardware, and construction materials.

By using COTS components, it is possible to create a cost-effective prototype of a particular design project. For example, Winchenbach and Segee (2011) point out that by acquiring and assembling COTS from the market, it is possible to reduce significant time and cost in designing a mobile robotic platform. The use of COTS to improve cataloging of Inner Earth Object (IEO) items was implemented by the German Aerospace Center in its AsteroidFinder mission. This approach allowed the development of an efficient and robust system design solution within the limitations of a smaller satellite project (Findlay et al., 2011). Several leading aerospace companies have started using new solutions employing COTS tool providers and in the process have discovered that these methods were the best fit for the individual needs of product developers (Low, 2011).

It is important for students to search for what COTS components are available that they can use in their design solution. While significant reduction in cost is possible by using a COTS approach, there are other issues such as reliability and quality that need to be considered. While searching for such components, focus on these issues is critically important in designing a product which is both reliable and cost-effective.

PROCEDURE OF MATERIAL SELECTION

Properly selecting materials is a critical step in determining the best solution for a design application. It must be noted that the process is typically not linear, since there are separate design requirements that depend on specified design criteria; it is not just the physical properties that determine the best material. For example, the budget will be set by the client, and the client may want the product to look a certain way for marketing purposes. These considerations must be taken into account throughout the selection process.

The first step is to determine the physical constraints on the design item, such as size, loads, and durability (see Box 12.1). Once these constraints have been determined, they are used as inputs that are plugged into functions to determine the material physical properties required, such as density, strength, and stiffness. This is an important step that can immediately eliminate many possible materials due to inappropriate performance characteristics that simply will not do the job. Material selection charts are very useful in isolating the range of materials that have the correct prescribed property profiles (Ashby, 2005). For example, the CES Selector software offered through Granta Design (http://www.grantadesign.com) can generate various charts

Refine Solution

BOX 12.1

Steps in the Materials Selection Process

1. Translate design requirements
2. Screen using constraints (i.e., eliminate materials that can't do the job)
3. Rank using objectives: find materials that do job the best
4. Seek documentation: research the history of top-ranked candidates (see if there are pitfalls, or track record of performance of the materials)

Data from Ashby, 2012.

that are helpful in comparing various material properties desired for the specific design. If a very strong lightweight material is desired, strength-to-density and Young's modulus-to-weight ratios are dominant material properties. If embodied energy and cost are also concerns, strength-to-relative-cost and strength-to-energy-content could be deciding factors. The charts provided by the CES Selector and other such software can be used during the material selection process to isolate the area identifying all possible materials that apply to the design solution (Ashby & Cebon, 2007). Examples of charts for a variety of materials, along with an in-depth explanation of the significance of each chart, are available at http://www.me.uprm.edu/vgoyal/inme4011/Online_inme4011/Topic2_MaterialSelection/AshbyCharts.pdf.

A list of materials that have the desired properties can be generated using material selection charts to eliminate materials that fall outside the various design constraints. Once the materials with the required physical properties have been located, candidate materials can be ranked using objectives specific to the application and desire of the client and designer, such as aesthetics, manufacturability, or environmental considerations. If a material does not look good, cannot be practically manufactured, or degrades over time because of environmental exposure, it will not be a good choice.

The final element in material selection is total cost. Material selection charts can be used to calculate the cost per unit mass, which can be fed into total cost estimates based on how much of the material is needed compared to that needed for alternative design solutions. This procedure allows the student to separate design constraints from desirable material properties before selecting the least cost material that will be best suited for the application. Students also may want to research the history of top-ranked candidates to see if there are pitfalls, or a track record of performance that may raise caveats or reinforce the choice of that material.

LOCATING INFORMATION ABOUT MATERIAL PROPERTIES

Mechanical properties of materials, such as fracture toughness, tensile strength, hardness, creep, and fatigue strength, are predictors of the way materials behave during the application of different types of stress (Stoloff, 2012). For example, suppose a design problem requires exploring mechanical properties of materials to understand how much deformation a material can withstand before breaking or how much resistance a material offers to fracture. In this case, ductility and toughness are two examples of mechanical properties which need to be explored. Other mechanical properties include elastic moduli, yield strength, tensile (ultimate) strength, compressive strength, fatigue endurance, and failure strength. While understanding these properties is important, it is equally important to learn how to find material properties using a variety of information resources and tools currently available. These properties can be found in subject-based online handbooks, such as the *Engineer's Handbook* (http://www. engineershandbook.com), and scientific reference works that libraries subscribe to such as Knovel and CRCNetBase. It is important that students become familiar with using these online resources, as the more they use them, the more likely they will be to use high-quality sources instead of more dubious open Web sources in their search for appropriate materials. In this case, being able to search through compiled data has no substitute in the open Web. It is also important for students to always check the library's reference section for handbooks that will contain much of the same information found online.

SELECTED SOURCES OF MATERIAL INFORMATION AND DATA

There are a number of resources available that provide access to property data of different materials.

ASM Materials Information Online

The ASM Materials Information database (http://products.asminternational.org/matinfo/index.jsp) contains the contents of the ASM Handbook series, among other content produced by ASM. It contains peer-reviewed, trusted information in every area of materials specialization. This series is the industry's best known and most comprehensive source of information on ferrous and nonferrous metals and materials technology.

CES Selector

CES Selector (http://www.grantadesign.com/products/ces) is a powerful software application that offers extensive materials property data, advanced graphical analysis, and specialized tools to support materials selection and substitution decisions. The CES Selector database allows students to create interactive charts as a function of different properties to assist in the selection of appropriate materials. It was developed for the education market, providing an intuitive graphical interface and hyperlinked definitions of properties throughout, to assist students in navigating the material information landscape.

Knovel

Knovel (http://www.knovel.com) provides electronic access to leading engineering reference handbooks, databases, and conference proceedings. It was the first publisher to extract data from handbooks, allowing the search for material properties across a wide variety of titles.

The Materials Project

The Materials Project (http://materialsproject.org) is an open science initiative that makes available a huge database of computed material properties. The Materials Project aims to reduce guesswork from materials design in a variety of applications, as experimental research can be targeted to the most promising compounds from computational datasets. Researchers will be able to data-mine scientific trends in material properties. By providing materials researchers with the information they need to design better, the Materials Project aims to accelerate innovation in materials research.

Refine Solution

Matweb

MatWeb's (http://www.matweb.com) searchable database of material properties includes data sheets of thermoplastic and thermoset polymers such as ABS, nylon, polycarbonate, polyester, polyethylene, and polypropylene; metals such as aluminum, cobalt, copper, lead, magnesium, nickel, steel, superalloys, titanium, and zinc alloys; ceramics; plus semiconductors, fibers, and other engineering materials.

NIST Data Gateway

The National Institute of Standards and Technology (NIST) Data Gateway (http://srdata.nist.gov/gateway/) provides easy access to many (currently over 80) of the NIST scientific and technical databases. These databases cover a broad range of substances and properties from many different scientific disciplines. The Gateway includes links to free online NIST data systems as well as to information in NIST PC databases available for purchase.

LOCATING COMMERCIAL OFF-THE-SHELF (COTS) COMPONENTS

There are many resources available online that can assist in sourcing the appropriate COTS equipment and manufacturers (see Box 12.2). Some suppliers focus on providing only specialized types of material such as software and electrical, mechanical, and construction materials. For example, a very common COTS item is a power supply. Many products require power in order to function, and it is more beneficial to the designer to choose a pre-manufactured power supply rather than to design and produce it from scratch. There are several caveats to be aware of using COTS, such as the possibility of a third-

> **BOX 12.2**
> **Selected Sources of Information for Commercial Off-the-Shelf Components**
> **General**
> www.Thomasnet.com
> www.globalsources.com
> **Electrical/Software**
> www.freetradezone.com
> www.allelectronics.com
> www.3csoftware.com
> www.adafruit.com
> www.sparkfun.com
> www.makershed.com
> **Mechanical**
> www.mcmaster.com

party component vendor's going out of business or dropping the support of a certain product. When using a COTS component, it is important to view the spec sheets to determine what specifications and tolerances the component has been built to and to ensure as objective a comparison between components as possible. Consulting product review sites can also help when choosing between components to see whether a particular community believes the components are really performing up to their specifications.

SUMMARY

The embodiment of a design concept in order to make it a practical reality demands finding the right material or identifying the most appropriate components that can meet the product requirements. Selection is not a simple process. It must be undertaken in a disciplined and methodical way, using a coherent search strategy. It sometimes requires trial and error, experimentation, and analysis of results before the most cost-effective, environmentally sound material selection process is complete. There are numerous online resources, handbooks, and selection software to aid in this process. However, these tools are only as good as the underlying strategy that the designer using them adopts.

SELECTED EXERCISES

Since in the design process students may be searching for properties throughout a course, a good introductory exercise may take the form of a sample project in the beginning of the term. The faculty member teaching the class collaborates with a liaison librarian and together they set up an assignment requiring students to select a material and search for properties for the project. The librarian provides instruction to show how properties are located or calculated. A research guide highlighting a number of useful sources will help students determine which sources are available for researching materials and material properties. Students work in small groups and search using the various tools and resources provided in the research guide. In consultation with the faculty member and liaison librarian, students identify candidate materials for their project. This search experience will be used as the basis for their project as the group continues to identify and experiment to find the right final materials.

Exercise 12.1

Ask students to imagine that they are preparing to design a wind farm near Atlantic City, New Jersey. The turbines will be designed for a salt air environment and constant exposure to ultraviolet (UV) radiation. What material properties will be most critical when designing the blades? Why?

Since windmill blades are essentially cantilever beams bending under wind pressure, both strength-to-weight and stiffness-to-weight ratios will be important design parameters. Material resistance to salt air corrosion and UV degradation will be important environmental concerns in the design process as well. Special coatings may be needed.

Exercise 12.2

Using the table feature in Microsoft Excel, have students brainstorm a list of possible materials based on the required physical properties for their project. Once they have the list of materials that will meet the physical requirements, have them start analyzing each material for the next criterion, such as environmental considerations and cost. Using the filter feature in the table, they can turn off all materials that are eliminated based on the next set of materials. They are left with only the materials that have not been eliminated showing, making it easier to rank and compare various materials. Have them repeat the process for each criterion until only the best candidates remain.

Exercise 12.3

Structural materials are usually selected based on their stiffness (resist deformation) and strength (will not fail). But we also desire that they be lightweight, especially in aircraft. Ask students what parameter best accomplishes these objectives, and where they would find that data.

ACKNOWLEDGMENTS

We appreciate the Drexel Smart House initiative and its contributions to this chapter about its experience with student projects in selecting proper materials during design and prototyping. Drexel Smart House is a student-led, multidisciplinary project to construct an urban

Refine Solution

home to serve as a living laboratory for exploring cutting edge design and technology. Participants conduct research and develop design solutions aimed at improving the quality of life in urban residential settings. The program supports student innovation through early-stage research and the development of prototypes or models, with the ultimate goal of launching strong research and development for commercialization and technology transfer activities.

REFERENCES

Ashby, M. F. (2005). *Materials selection in mechanical design* (3rd ed.). Burlington, MA: Butterworth-Heinemann.

Ashby, M. F. (2011a). Engineering materials and their properties. In *Materials selection in mechanical design* (pp. 3–5). Burlington, MA: Butterworth-Heinemann.

Ashby, M. F. (2011b). Introduction—materials in design. The evolution of materials in products. *Materials selection in mechanical design* (pp. 36–37). Burlington, MA: Butterworth-Heinemann.

Ashby, M. F., & Cebon, D. (2007). *Teaching engineering materials: The CES EduPack*. Cambridge, UK: Granta Material Inspiration. http://stuff.mit.edu/afs/athena/course/3/3.225/refs/Teaching_Engineering_Materials.pdf

Ashby, M., & Jones, D. (2012). *Engineering materials 1: An introduction to properties, applications, and design* (4th ed). Burlington, MA: Butterworth-Heinemann.

Chopra, A. (2012). *Dynamics of structures: Theory and applications to earthquake engineering*. Upper Saddle River, NJ: Prentice Hall.

Dowling, N. (2007). *Mechanical behavior of materials: Engineering methods for deformation, fracture, and fatigue*. Upper Saddle River, NJ: Prentice Hall.

Farr, J. V. (2011). Costing and managing off-the-shelf systems. In *Systems life cycle costing: Economic analysis, estimation, and management* (pp. 207–230). Boca Raton, FL: CRC Press.

Findlay, R., Essmann, O., Müller, H., Pedersen, J., Hoffmann, H., Messina, G., & Mottola, S. (2011). AsteroidFinder: Implementing a small satellite mission to detect IEOs. In *62nd International Astronautical Congress 2011, IAC 2011*, 5, 3678–3687, International Astronautical Federation, IAF.

Ho, L. (2012, December). *Investigators report on probable cause for giant Chinese shark tank failure*. Advanced Aquarist. Pomacanthus Publications, LLC. Retrieved from http://www.advancedaquarist.com/blog/investigators-report-on-probable-cause-for-giant-chinese-shark-tank-failure

Jahan, A., & Edwards, K. L. (2013). Weighting of dependent and target-based criteria for optimal decision-making in materials selection process: Biomedical applications. *Materials & Design, 49*, 1000–1008. http://dx.doi.org/10.1016/j.matdes.2013.02.064

Low, J. (2011). Platform engineering approach to the electrical systems architecture development process. *SAE International Journal of Aerospace. 4*(2), 944–951.

Masuelli, M. A. (2013). Introduction of fibre-reinforced polymers—Polymers and composites: Concepts, properties and processes. In M. A. Masuelli (Ed.), *Fiber reinforced polymers—The technology applied for concrete repair*. New York: InTech. http://dx.doi.org/10.5772/54629

Stoloff, N. S. (2012). Metal, mechanical properties of. In *AccessScience*. Retrieved from http://www.accessscience.com/content.aspx?searchStr=mechanical+properties&id=417810

Winchenbach, S. A., & Segee, B. (2011). A cost-effective mobile robot platform using commercially available components. *Journal of Computer and Information Technology, 1*(3), 2–8. Retrieved from http://www.academypublish.org/papers/fullpdf/4v1n3.pdf

CHAPTER 13

GET YOUR MESSAGE ACROSS

The Art of Gathering and Sharing Information

Patrice Buzzanell, Purdue University

Carla Zoltowski, Purdue University

Learning Objectives

So that you can guide students in improving their professional communication skills and develop more persuasive presentations, upon reading this chapter you should be able to

- Identify common challenges to successful communication in different kinds of presentations
- Describe how to map a process for designing effective presentations
- Describe strategies for identifying the most critical information to communicate to stakeholders
- Outline ways students can identify likely responses to their presentations so that they can anticipate and address those questions
- Evaluate how using different media may enable students to achieve their presentational goals more efficiently

INTRODUCTION

Typically, the student design project culminates with a formal presentation and written documentation given to the instructor and clients or other stakeholders of the project. This is the opportunity for the students to demonstrate what they have learned and achieved in the course of their project, and showcase their skill in distilling this knowledge so that they provide the essential, relevant information in a concise, coherent, and persuasive manner.

Although the final presentation is the dominant focus when students think about communication, throughout the engineering design process, there are multiple opportunities to communicate with various stakeholders who have a vested interest in particular design processes and outcomes. Chapter 7 describes active information gathering techniques that enable presenters to obtain relevant design information. This chapter on effective communication with stakeholders discusses how to convert stakeholder information as well as other parts of the design process into talking points within an effective presentation.

These opportunities enable designers to listen for and be responsive to stakeholders' real interests and not simply what they state that they need. These opportunities involve information and opinion seeking for the necessary details to fulfill criteria for design specifications, to acquire resources for prototype development, to assess the quality of prototypes, and to sustain the viability of deliverables. In short, the steps for effective communication with stakeholders begin long before designers face their final presentations. However, it is in these final presentations that designers want to persuade stakeholders to accept particular solutions. The satisfactory outcomes of such presentations are not simply agreement about implementations, but also maintenance of good working relationships among key stakeholders and mutual respect for different types of knowledge that each brings to bear upon the design solution.

In this chapter we define communication as the ability to articulate—through speech, written texts, and graphic representations—different stakeholder interests and design considerations for team deliberations and public presentations. To achieve good communication in general and persuasive ability in particular, it is necessary to recognize what is needed and competently perform the spoken, written, and/or graphic presentations. Competent presentations take into account the diversity among stakeholders and variety of formats, including one on one, team based, in person, and virtual. It is also necessary to recognize that for different design phases and stakeholders, different levels of technical detail are preferable. Finally, there are specific argument formats that typically are effective in persuading other team members and external stakeholders as to the efficacy of design decisions and solutions.

COMMON CHALLENGES FOR STUDENTS

In this section we identify several common challenges to successful communication in presentations. When presenters can identify which challenges are applicable to their specific presentational goals and contexts, they are able to focus their attention on what they need to work on the most. Doing so enables them to make good use of their time as they work toward effective presentations.

The first challenge is to realize that not everyone understands the big picture of the design project. Another way of phrasing this challenge is: What is the story that presenters want to tell? What do presenters want audience members to know, feel, and/or do at the conclusion of the presentation? Often students focus on the details or aspects that are most salient to them at the time and tend to not step back and translate the big picture story for their particular audience (Dannels, 2002, 2009; see also Gallo, 2009). This first challenge is particularly difficult because it requires flexibility in thought and ease with presenting both macro and micro issues involved in the design process and proposed solutions.

One way to work on this first challenge is to provide a short history of the project. When did the project begin? What was the motivation for the project? (For example, what device, tool, or process is the client currently using?) What goal or end are you trying to achieve with the project? Who are the stakeholders? What is the context of the project? Why and how was the design team assembled? Supplying this information at the beginning of the presentation provides the audience with the context that is often needed to understand the design criteria and justifications provided in the remainder of the presentation.

A second challenge is knowing the audience for the presentation as well as what kinds of arguments and information are relevant to that audience. For example, a presentation to end users would focus more on characteristics of the design solution as related to their needs, whereas a design review presentation to clients would include more technical design solutions and explain why certain design decisions were made. In knowing who the audience likely will be and what their vested interests are, the presenters can address exactly what key points audience members would want to know. Some might want to know how the proposed design solu-

tion would work, or how much it would cost to develop a feasible prototype. Others might be concerned about training personnel and safety issues. When audience members are operating in a business model, financials become more relevant than when audience members primarily work for nonprofits, where values and client service are priorities. In a business or entrepreneurship setting, it often is important to present a detailed budget and to anticipate questions about line items. The consequences of budget projections would be prominent in these audience members' minds. If the team cannot argue that there is a benefit (or decreased cost), then the design solution would not be acceptable to some audience members. In sum, knowing the audience helps the team to not only construct a presentation that meets audience members' informational needs but also anticipate audience members' responses.

A third challenge is to construct a presentation that would be considered well organized by audience members. Although an introduction-body-conclusion format works well for informational presentations, there are other structures that are advantageous if the goal is persuading audience members to change their thinking or behavior. One such format is a problem-solution format in which presenters first sell audience members on their version of what the problem is and provide evidence that supports their particular problem statement(s) (Beebe, Beebe, & Ivy, 2008). Once audience members understand and buy into identification of the problem, then possible solutions are presented along with the extent to which each solution satisfies the problem specifications. Once alternatives are eliminated, then audience members should readily agree to the proposed solution. Of importance to the organization of the presentation is that presenters know what kind of format would be both easy

Communicate Effectively

to follow by audience members and fulfill the presentational goals.

The fourth challenge is demonstrating credibility or trustworthiness. The response to this challenge begins early in the design process when the team does an assessment of what knowledge, skills, and abilities (KSA; see Hartenian, 2003) are essential to project problem identification and solutions. Periodically, the team will consider other needed KSAs and determine how such individual competencies are shared to improve team effectiveness (Delamare Le Deist & Winterton, 2005; Littlepage, Perdue, & Fuller, 2012).

When KSAs are presented to audience members, these audience members will understand how the team was composed. Moreover, the KSAs operate as areas on which team members can build credibility as they present the research they had conducted and the specialists with whom they consulted. The challenge is not simply listing KSAs but showing how team members' KSAs were used to design an optimal solution.

Presenters' credibility is greatly enhanced when they can speak firsthand about conversations they have had with clients, potential users of the design product, and others who have vested interest in the solution. A challenge during presentations that involve technical and engineering personnel is to relay points with enough technical detail for some audience members without losing others who are more interested in other aspects of the presentation.

A final consideration for the challenge of demonstrating credibility is presenters' response to questions by audience members. An ability to provide further explanation to questions is very important and can be practiced so that students are well prepared. It is also important to respond appropriately to questions for which they do not know the answers.

Sometimes, when presenters do not know the answers, they might make up answers instead of saying "I don't know." Therefore, demonstrating credibility also means admitting that there are design aspects that team members did not consider and/or questions to which they do not know the answers, but can explore further.

In sum, design presentations involve a number of challenges. However, some of the most common challenges are telling the story, knowing the audience, organizing the presentation effectively, and displaying credibility without losing audience members who do not share the same level or kind of KSAs. In short, when presenters provide insight into how and by what criteria decisions are made—with documentation—and involve the stakeholders, then they are presenting with integrity.

PERSUASION WITH INTEGRITY THROUGHOUT THE DESIGN PROCESS

Because persuasion occurs throughout design processes, the groundwork for selling solutions has been laid from the very first connections among team members and stakeholders. The goal is not simply to develop a presentation that encourages decision makers to accept a particular solution, a common definition of *persuasion*, but also to create knowledge with all stakeholders throughout the design process so that the solution under discussion is neither a surprise nor unworkable. Furthermore, persuasion typically involves attempts to enable stakeholders to exercise choice among various ways of thinking, knowing, and feeling about information and design features such that their behaviors in approving or modifying design solutions are accomplished. These characteristics of persuasion mean that persuasion is a process involving information literacy and the understanding of human nature. These features also mean that informed choices, rather than coercion or unethical arguments, can produce the best solutions at any point in the design process. Although these characteristics make sense for effective persuasion and design, without exception we hear our engineering design students voicing frustrations that they "can't get other team members to do what they want," thus failing to recognize the process-oriented nature of persuasion and the need to know the interests, knowledge levels, disciplinary concerns, and emotional connections to the project that team members (and other stakeholders) hold.

Although stakeholders may change during the course of a project, designers can anticipate and prepare for the unique challenges and opportunities in selling solutions to different stakeholders by mapping out the design process with both the necessary communication and technical knowledge running parallel.

IDENTIFY CRITICAL INFORMATION TO COMMUNICATE

Many different categories of criteria are considered when developing a design solution: functional performance, form, aesthetic, economic, environmental, ethical, health and safety, inclusiveness, manufacturability, political, social, sustainability, and usability. In determining what information is critical to communicate, seasoned designers recognize that in design and any kind of persuasive activity there are conflicts because choices made at every step are not made without some trade-offs between different criteria, and that individual audiences and disciplines prioritize them differently. Some in-

terests are fairly predictable. For instance, engineers are interested in safety and human costs, compared to the features and aesthetics that might be of interest to industrial designers and architects, or the feasibility of design and cost factors that might gain building and construction specialists' notice. These are general disciplinary or occupational patterns that designers can anticipate as priorities for their audiences.

Sometimes designers or others involved in persuasion fail to realize that people have different priorities because of their interests, jobs, and values. Researchers, such as Paul Leonardi (2011; see also Barley, Leonardi, & Bailey, 2012) as well as Carrie Dossick and Gina Neff (2011), have examined how members of multidisciplinary engineering design teams work together to persuade each other and different stakeholders about their viewpoints concerning design outcome or deliverables. These authors examine multiple phases in engineering design as well as the communication among different stakeholders with varied interests in the deliverables. They recognize not only that engineering design and multidisciplinary collaborations in general are messy because certain disciplinary interests or logics, such as safety for engineers, sometimes override other concerns, but also that problem definition and criteria for alternative and prototype design become complicated when there are diverse vested interests and disciplinary jargon. As a result, a substantial amount of time needs to be budgeted to work through (sometimes) unpredictable communication with stakeholders. Another important consideration for effective design solutions and their presentation is that clarity is not simply a written or oral feature in language choices and presentational format but also requires the selection of material objects. These material objects may include sketches, YouTube presentations, graphs, charts, computer-aided design (CAD), software code, and prototypes.

There also may be incidents reported in the news that raise awareness or concerns relative to the project design. The function of persuasion in these disciplinary and newsworthy cases might be to encourage different stakeholders to negotiate and reframe the evaluation of certain criteria over others at particular design phases. In the presentation where the final design deliverable is submitted for stakeholder approval, discussions about such considerations and their negotiation should be reported. Acknowledging the shifts in decision-making criteria throughout the design process enables audience members to revisit their previous concerns and how presenters have incorporated this feedback into their solutions. In these ways, designers legitimize stakeholders' disciplinary, newsworthy, or other concerns and focus attention on the processes that led to the solution.

PACKAGE CRITICAL INFORMATION FOR SUCCESSFUL PRESENTATIONS

To determine critical information to communicate, especially in design review presentations where the goal is to secure stakeholder support for design decisions and process, designers can be guided by some standard criteria. Design valuators typically look for (a) problems and context, (b) design fixation, (c) measurable ways to meet design specifications, and (d) specificity and verifiability.

First, when persuading others, evaluators want to know about the *problems and context* in which deliverables are going to be used. Those making decisions want to know that designers understand not only who the potential users of the design solution are but also how that solution fits within these stakeholders' and un-

anticipated users' lives. By indicating that they are well aware of the problems driving particular designs, designers communicate depth and breadth of knowledge. Therefore, presentations should include the following:

1. When did project begin (overall timeline)?
2. What was the motivation for the project? (For example, what device, tool, or process is the client/user currently using?)
3. What is the project goal or end?
4. Who are the stakeholders?
5. What is the context of the project?

For instance, during one design team presentation, the members did not provide enough contextual information or their vision for the ways that their design solution would meet po-

Smart Goals

Ideally, designers present their project goals in ways that their evaluators can readily assess whether or not the project is appropriate. There are many ways to construct presentations, but SMART (specific, measurable, attainable, realistic, and timely) project and customer requirements and specifications provide some ready criteria. These criteria ask designers to respond to anticipated questions in areas already covered: What did designers consider, whom did they involve, and how did they make decisions? What assumptions are being made? From where did the requirement come? How will designers know when they have met the specifications and requirements? Have the specifications and requirements been met? The responses to these questions provide insight into decision making and design process and are critical for evaluators to appropriately assess the design solution. These anticipated questions also increase the chances that designers will obtain appropriate feedback for their goals.

Communicate Effectively

tential users' needs. Design evaluators provided detailed feedback for a high action soccer game in which players' kicking skills needed to be further developed through exercises and equipment. What the team failed to convey to their evaluators was that the soccer-assist project was developed for children with special needs who required modifications in standard exercises, equipment, and so forth. As a result, the designers missed an opportunity to obtain useful and appropriate feedback about their processes. However, they did learn a lesson in framing their project vision and mission at the outset of their presentation. They learned how to present the problems that they were facing through detailed scenarios and video-recorded segments. In short, they showed design evaluators how the problems and context required that they learn more about the capabilities of their potential users.

Second, design evaluators look for instances of *design fixation*, a process by which engineering design team members become committed to a particular design solution to the extent that they may no longer listen to and process information that contradicts or expands their original solution. Design fixation is more common among novice designers rather than experts, who are better versed in the fluidity of design processes and knowledge creation (Crismond & Adams, 2012; Cross, 2000; Gero, 2011).

When evaluators see that designers want to focus solely on solutions rather than the problems, they become suspicious. Focusing on solutions might indicate that designers are hiding or are unaware of problems. These quick fix solutions may indicate that designers simply want to sell their solutions or that they are engaged in design fixation. Designer evaluators might ask directly or imply that they have concerns: In whose interests were particular solutions designed? Why does critical thinking seem to be missing from the design processes? Why do the

> ### Assessing and Communicating Risk
> DFMEA (design for failure mode effects analysis) is a useful tool for identifying potential sources of failure; evaluating the occurrence, severity, and ability to detect the risk; and anticipating likely outcomes of the design solution and previously unanticipated considerations that might prove detrimental to users. These risk considerations and evaluations speak to design processes in general as well as to issues that should be raised or considered when communicating solutions. Designers need to present information that indicates that they have considered risk. This information may include materials that add credibility to the design process itself—photos, sketches, modeling, and simulations for prediction of different outcomes—as well as to the information presented and source credibility.

data not match the rest of the presentation (i.e., lying with data or constructing claims based on little or no data)? How have designers assessed risk? Once designers' credibility has been questioned, it is difficult to rebuild trust. As mentioned earlier, insight into the decision making throughout the process and at particular times or milestones can lessen evaluators' concerns (e.g., Buzzanell, in press).

Third, evaluators want to learn how design team members are able to meet design challenges, that is, to be presented with *measurable ways to meet design criteria*. As noted above, designers need to present data indicating thorough analysis of the context and problem so that the design solution seems not only reasonable but optimal. In linking data with solutions, designers address the following:

1. Feasibility (that they have or know where to locate technical capacities to fulfill the solution)
2. Desirability (that there is a human need or desire for the solution)

Communicate Effectively

3. Viability (that the solution is economically possible and sustainable) (Brown, 2009)

Presenting measurable ways to meet design specifications also indicates that the designers understand the process and admit times when their decision-making phases required that they obtain additional feedback or they took a wrong turn. Such detailed information requires that individual and team documentation be specific and verifiable—that is, include enough detail, data, and sources such that design evaluators feel as though they can readily check into the truth of claims and solutions.

Fourth, although *specificity and verifiability* seem fairly obvious ways to build credibility for selling solutions (see Rosenthal, 1971), they are more difficult than they first seem. Not only do these processes require documentation at every design phase that can be readily accessible for information support in the selling-your-solutions presentations, but also they require that presenters be perceived as credible or trustworthy and ethical.

How do designers know if design evaluators or other stakeholders will see the quality of their information and themselves as specific and verifiable? As mentioned earlier in this chapter as well as in earlier chapters in this handbook, these qualities result from an analysis of stakeholders to figure out what they need to assess information as specific and verifiable. For the soccer-assist project we described under the problems and context criterion for effective presentations that designers might expect (and should verify) that community members—business owners contributing funds, parents of children with special needs, and others—would be less interested in the detailed reports about the engineering principles underlying potential design solutions than about how their own or neighbors' children might use

safe equipment. They may be less interested in a technical article in an academic journal that they have never heard of than in a summary of key issues relevant to the the soccer-assist project design solution that comes from the same journal, published within the current year, and deemed highly credible because of designers' commentary that it is the premier academic journal in the area and one on which sports, physical, and occupational therapists rely. Key stakeholders would learn about the solution details that meet specifications and the prestige and usefulness of sources from which such decisions resulted. They would know what to look for and where such information could be obtained—meaning that they are more likely to accept solutions being presented without checking into these details because they believe such information is trustworthy.

For engineering and other technical or specialized audiences, further details including schematics, technical jargon, and additional academic sources enhance perceptions that designers did their homework and can be trusted to accurately portray the bases on which solutions are derived. Specificity and verifiability also refer to presenters' credibility. Stakeholders want to know why and how designers are interested in and might have conflicts of interest with particular problems and solutions, including self-references indicating personal interest, experience, or loyalties in an area. Prestige references or referral to well-regarded sources (e.g., academic journals ranked best in quality, business or disciplinary newsletters held in high esteem, people whom stakeholders know and trust) aid designers in selling their solutions. For the soccer-assist project, designers who have played soccer, worked with or have children with special needs in their friendship and family circles, or who have focused their career on designing for individuals with special needs would have more credibility with their statements about such interests and background inserted at appropriate times during presentations. These self-references and prestige references need not be detailed but they are powerful.

KNOW HOW THE AUDIENCE VIEWS YOUR PRESENTATIONS

The sections we have covered thus far in this chapter have focused on understanding and managing design evaluators' interpretations, informational needs, and expectations. In a nutshell, they require that designers persuade others to a particular understanding of the problem and to a solution that meets design criteria specified in the previous section.

Persuading others is dependent not only on the designers, or sources of problem and solution presentations, but also on those who evaluate and must live with design solutions. As a result, it is insufficient to learn techniques for persuading others without learning how messages might be processed.

In general, people process both habitually, using *heuristics*, and mindfully, using more active cognitive processing. Heuristics, or heuristic principles, "represent relatively simple decision procedures requiring little information processing" (O'Keefe, 2002, p. 148). Varieties of heuristic principles include credibility, liking, and consensus. We actually have talked about heuristics when we mentioned that specificity and verifiability in information and provided by designers can enhance the chances that design evaluators and other stakeholders will accept solutions rather than digging for more information or questioning feasibility, desirability, or viability. For credibility, highly

trustworthy and effective presenters are those who provide enough information, tailored to audience interests and knowledge, delineating assumptions and risks, and embedded within the context. Such credible presentations are enhanced if design evaluators like or respect the presenters (known as the *liking heuristic*) and if designers can state truthfully that others have reviewed and approved the solution (known as the *consensus heuristic*). These heuristics do not mean that presenters need to be friends with design evaluators or detail every single approval step, but that presenters seem approachable, eager to explain their processes, and willing to answer questions and/or admit that they are human (i.e., perhaps have not considered every possible angle or question).

In addition, we assume that presentations of self, design processes, solutions, and context would be truthful and enthusiastic. We also assume that arguments and evidence would be well organized, data rich, and results oriented (see Dannels, 2002). Overall, then, effective presentations frame desirable interpretations of information and construct the knowledge structures in which design evaluators can make decisions about the content and presenters themselves. Persuasion can come about through these peripheral processes.

Rarely, however, are design solutions processed habitually with such simple decision rules or principles. The chances of heuristic processing happening are increased when designers have sought information and opinions throughout their design processes—meaning that when they are selling their solution, they have already countered objections and have utilized and credited their previous sources for their information. At these times, evaluators may use peripheral or heuristic processing because they are unmotivated to engage more actively (i.e., to them, design criteria have been met by the solution).

Communicate Effectively

More likely, designs are reviewed mindfully, meaning that design evaluators fall somewhere between heuristic or peripheral processing and more *active cognitive processing* in order to process the information (see Gass & Seiter, 2009; O'Keefe, 2002). Active cognitive processing occurs when audience members do not simply accept solutions but ask questions, incorporate their own information, assess solutions critically, and generate their own alternatives and optimal solutions. Given that new evaluators and stakeholders may enter the design process at any point, it is useful to always be prepared for active or central processing. To prepare for active cognitive processing, designers should engage in one or more trial runs of the presentation. During this trial run, high-quality arguments—specific and verifiable—should be offered with precise definitions and support. Not all of the information for which designers prepare will be used for the actual presentation. The detailed criteria, sources, and findings about contexts and problems would be available in a separate presentation section (after the closing and question-answer phase of the presentation) or in a different PowerPoint presentation and other documentation (see Schoeneborn, in press). Practice during trial runs and preparation of supporting materials are particularly valuable for face-to-face and online design critiques in which stakeholders often provide feedback based both on the relationship that they have developed with the designers and on particular questions or recommendations that they would like to pose (Dannels, 2009, 2011).

The point is that these answers to questions and objections to the solution that is being offered are available for review. It comes down to a tradeoff—presenting just enough information in a readily accessible format without going overboard and without underestimating evaluators' questions and concerns.

USE MEDIA EFFECTIVELY

Media and material objects enable designers to distill information from multiple sources and communicate it appropriately, ethically, and credibly. A segment from a video depicting rural village life in Ghana can provide more information about the context, major stakeholders, problems, and specifications than can an elaborate speech. Likewise, engineers on multidisciplinary teams use material objects, such as sketches, drawings, photos, CAD models, and so forth, to explain what they mean quickly and easily. In using any media, the criteria for inclusion are as follows: How can incorporation of these media or objects move design evaluators toward accepting the solution being presented? Do these media or objects help build support for feasibility, desirability, and viability? Are there potential questions about the media or objects that presenters cannot answer or that divert attention from the primary presentational goal—namely, selling a solution? Finally, do the media or objects add to clarity, elaborate on key points, or bolster presenters' credibility in some way? For instance, Skyping with partners from a Ghanaian water energy education initiative or capturing their voices and videos ahead of time can do more to indicate designers' commitment and credibility as well as the context than all the words in the world!

SUMMARY

In this chapter we presented some key considerations in constructing effective design presentations and in anticipating audience members' responses. Students need to mine the information they have gathered throughout their design process, including stakeholder needs, alternative solutions considered, and the performance of the anticipated design deliverable, and distill the information that will be most important to their audiences. This will enable them to make the best use of their time with the clients so that their core message will have enough support to be persuasive without being too weighed down with details and therefore obscured. In this chapter we identified common challenges, presentational design processes, strategies to identify and use critical information, and ways to anticipate stakeholders' interests and concerns.

Although the final presentation is frequently the culminating activity in a design project, much daily engineering and multidisciplinary teamwork is done in interpersonal and group experiences (Darling & Dannels, 2003). Throughout their projects, students should be encouraged to communicate frequently with clients and other stakeholders in order to create a shared understanding of the desired outcomes so that the final presentation is not a shock to either side, but rather the final step in a logical conversation.

SELECTED EXERCISES

Exercise 13.1

Break students into their design teams and have them identify the most critical information to communicate to each of the stakeholders of their design project. Ask them to anticipate questions the different stakeholders may have and how the design team might respond. Have each design team share strategies for meeting the information needs of their stakeholders.

Exercise 13.2

Break students into their design teams and have them brainstorm different media that would enable them to meet their presentation goals and encourage design evaluators and participants in the presentation to engage with the materials.

Exercise 13.3

Have students map a process for designing effective presentations, perhaps treating the presentation as a mini-design process itself. Have students describe common challenges in putting together an effective presentation.

ACKNOWLEDGMENTS

We would like to thank the Engineering Information Foundation (EIF) for support of our work in exploring communication strategies within student design teams.

REFERENCES

Barley, W. C., Leonardi, P. M., & Bailey, D. E. (2012). Engineering objects for collaboration: Strategies of ambiguity and clarity at knowledge boundaries. *Human Communication Research*, *38*(3), 280–308. http://dx.doi.org/10.1111/j.1468-2958.2012.01430.x

Beebe, S., Beebe, S., & Ivy, D. (2008). *Communication: Principles for a lifetime, portable edition—Volume 4: Presentational speaking*. Upper Saddle River, NJ: Pearson.

Brown, T. (2009). *Change by design: How design thinking transforms organizations and inspires innovation*. New York: HarperCollins.

Buzzanell, P. M. (in press). Reflections on global

Communicate Effectively

engineering design and intercultural competence: The case of Ghana. In X. Dai & G. M. Chen (Eds.), *Intercultural communication competence: Conceptualization and its development in contexts and interactions*. Newcastle upon Tyne, UK: Cambridge Scholars Publishing.

Crismond, D. P., & Adams, R. S. (2012). The informed design teaching and learning matrix. *Journal of Engineering Education*, *101*(4), 738–797. http://dx.doi.org/10.1002/j.2168-9830.2012.tb01127.x

Cross, N. (2000). *Engineering design methods: Strategies for product design* (3rd ed.). New York: John Wiley & Sons.

Dannels, D. (2002). Communication across the curriculum and in the disciplines: Speaking in engineering. *Communication Education*, *51*(3), 254–268. http://dx.doi.org/10.1080/03634520216513

Dannels, D. P. (2009). Features of success in engineering design presentations: A call for relational genre knowledge. *Journal of Business and Technical Communication*, *23*(4), 399–427. http://dx.doi.org/10.1177/1050651909338790

Dannels, D. P. (2011). Relational genre knowledge and the online design critique: Relational authenticity in preprofessional genre learning. *Journal of Business and Technical Communication*, *25*(1), 3–35. http://dx.doi.org/10.1177/1050651910380371

Darling, A. L., & Dannels, D. P. (2003). Practicing engineers talk about the importance of talk: A report on the role of oral communication in the workplace. *Communication Education*, *52*(1), 1–16. http://dx.doi.org/10.1080/03634520302457

Delamare Le Deist, F., & Winterton, J. (2005). What is competence? *Human Resource Development International*, *8*(1), 27–46. http://dx.doi.org/10.1080/1367886042000338227

Dossick, C. S., & Neff, G. (2011). Messy talk and clean technology: Communication, problem-solving and collaboration using Building Information Modelling. *Engineering Project Organization Journal*, *1*(2), 83–93. http://dx.doi.org/10.1080/21573727.2011.569929

Gallo, C. (2009). *The presentation secrets of Steve Jobs: How to be insanely great in front of any audience*. New York, NY: McGraw-Hill.

Gass, R., & Seiter, J. (2009). Persuasion and compliance gaining. In W. F. Eadie (Ed.), *21st century communication* (pp. 156–164). Thousand Oaks, CA: Sage.

Gero, J. S. (2011). Fixation and commitment while designing and its measurement. *The Journal of Creative Behavior*, *45*(2), 108–115. http://dx.doi.org/10.1002/j.2162-6057.2011.tb01090.x

Hartenian, L. (2003). Team member acquisition of team knowledge, skills, and abilities. *Team Performance Management*, *9*(1/2), 23–30. http://ds.doi.org//10.1108/13527590310468033

Leonardi, P. M. (2011). Innovation blindness: Culture, frames, and cross-boundary problem construction in the development of new technology concepts. *Organization Science*, *22*(2), 347–369. http://dx.doi.org/10.1287/orsc.1100.0529

Littlepage, G., Perdue, E. B., & Fuller, D. K. (2012). Choice of information to discuss: Effects of objective validity and social validity. *Small Group Research*, *43*(3), 252–274. http://dx.doi.org/10.1177/1046496411435419

O'Keefe, D. (2002). *Persuasion: Theory & research* (2nd ed.). Thousand Oaks, CA: Sage.

Rosenthal, P. I. (1971). Specificity, verifiability, and message credibility. *Quarterly Journal of Speech*, *57*(4), 393–401. http://dx.doi.org/10.1080/00335637109383084

Schoeneborn, D. (in press). The pervasive power of PowerPoint: How a genre of professional communication permeates organizational communication. *Organization Studies*. http://dx.doi.org/10.1177/0170840613485843

Communicate Effectively

CHAPTER 14

REFLECT AND LEARN

Capturing New Design and Process Knowledge

David Radcliffe, Purdue University

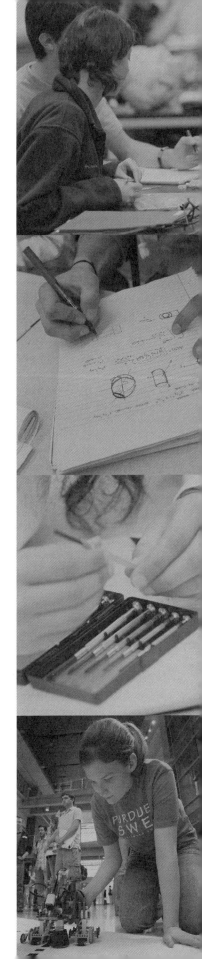

Learning Objectives

So that you can guide student teams on effective strategies for extracting deep learning from their design projects, upon reading this chapter you should be able to assist them to

- Choose a disciplined framework for reflecting on their practice as a means to learn and improve

- Capture and appropriately document design information and knowledge generated during a project

- Systematically capture the lessons learned about the process of team-based design

INTRODUCTION

Frequently, students, instructors, and indeed practicing engineers view the final presentation and documentation as the end of a design experience. However, the lessons are not fully learned until students have reflected on their experiences and internalized their insights into their professional practice. Engineers tend to be results oriented. They focus on solving a problem and once it is solved, and the challenge is over, they move on to the next project. However, during the course of any design project new technical knowledge is created and the teams can learn important lessons about how to work as a team in such a project. This new knowledge and lessons on process can usefully be applied to future projects so as to avoid reinventing the wheel or suffering the same frustrations in not having a team perform well for the same reasons over and over again.

Unfortunately, experience from engineering practice in many industry sectors suggests that too often this knowledge is not adequately extracted, articulated, captured, and/or transferred to future projects. Large engineering organizations have knowledge management systems that are designed to overcome this shortcoming, but the lesson learned database is often only sparsely populated or even empty. Often it only gets sufficient attention after there is a major failure (see Boxes 14.1 and 14.2).

Whereas once such knowledge management systems were paper based, now they take the form of sophisticated computer-based systems. Just as libraries have moved toward more digital repositories, so it is with lessons learned databases. However, this change in the technology of storage and indexing has not changed the tendency of engineers to do a very basic job of documenting the outcome of a project, beyond that necessary to meet contractual requirements.

> ## BOX 14.1
> ### NASA Lessons Learned Database
> Following the loss of NASA's space shuttle *Challenger* and crew in 1985, the NASA Lessons Learned program was formulated to assure that NASA's key knowledge is documented and made available to everyone, both the public and NASA personnel.
>
> Following the loss of NASA's space shuttle *Columbia* and crew in 2003, the Columbia Accident Investigation Board was convened to identify underlying causes of the accident. The Board determined that NASA's organizational structure and culture prevented it from being a learning organization. One proposed solution to this problem was the NASA Engineering Network (NEN), a suite of information retrieval and knowledge-sharing tools aimed at facilitating communication among engineers at all the NASA centers and affiliated contractors, thus taking knowledge sharing from availability to participation and collaboration.
>
> From NASA, 2010.

To the extent that engineering design is a learning activity, the design cycle is not fully closed (see Figure 1.3) until reflection has occurred to extract meaning, generate new ideas, or improve design processes. As discussed in Chapter 2, the *How Students Learn* report (National Research Council, 2005) advises that effective learning requires students to address their preconceptions (and overcome misconceptions), develop competence through a conceptual framework to organize the knowledge they have developed, and take ownership of their learning process, including developing skills to monitor their own progress and competency level.

Although reflection and knowledge management principles should be integrated throughout the design process, as indicated in the Information-Rich Engineering Design (I-RED) model, the culmination of a project provides the final opportunity to reflect on the entire process, allowing students to extract more global learnings about the project and their and their teammates' participation in it, as well as aggregate the reflections and learnings they gathered throughout the process.

COMMON CHALLENGES FOR STUDENTS

Perhaps not surprisingly, engineering students anticipate the behavior of engineers in practice in that they tend not to take the time to reflect in a disciplined way on projects they undertake in order to extract lessons and learning that can be transferred to future work. This natural disposition is reinforced when grades are assigned predominately on the basis of the technical deliverables in student design projects.

An increasing number of universities and colleges include critical thinking as one of a set of core learning outcomes (or competencies upon graduation) for engineering (and other) students. Unfortunately, the operational reality is that many of these schools do not integrate intentional learning activities into courses and curricula designed to develop and explicitly reward practices such as disciplined reflection

---

BOX 14.2

Lessons Learned: Information Systems Must Be User Friendly

Following the failures of the Mars Climate Orbiter and Mars Polar Lander in the late 1990s, the Office of the Chief Engineer was tasked with developing a plan for implementing the resulting mishap investigation boards' recommendations. The Office's report, released in 2000, made the following observations relating to lessons learned.

> The continuous capture and application of project knowledge and lessons learned must become a core business process within the Agency's program and project management environment. Regular input into NASA's knowledge bases, such as the lessons learned database, should be emphasized. Programs and projects should implement a "document-as-you-go" philosophy, promoting continuous knowledge capture for the benefit of current and future missions. More importantly, program and project managers must regularly utilize the knowledge management tools to apply previous lessons learned to their own projects. The Agency can provide help for individuals to understand, learn from, and apply the lessons of others to their own work as part of a daily routine.

> As of January 2012, the Agency has not met those goals. In fact, NASA's Aerospace Safety Advisory Panel recently stated in its 2011 Annual Report that in spite of excellent examples of individual and specific programmatic efforts to facilitate knowledge sharing, these efforts do not ensure the identification and capture of critical knowledge or provide for an Agency-wide single process or tool for locating and accessing all information resources.

> Specifically, we found that LLIS is underutilized and has been marginalized in favor of other knowledge management tools such as *Ask Magazine* and the annual Project Management Challenge seminar. Users told us they found LLIS outdated, not user friendly, and generally unhelpful, and the Chief Engineer acknowledged that the system is not operating as originally designed. Although we believe that capturing and making available lessons learned is an important component of any knowledge management system, we found that, as currently structured, LLIS is not an effective tool for doing so. Consequently, we question whether the three quarters of a million dollars NASA spends annually on LLIS activities constitutes a prudent investment.

From Office of Inspector General, 2012.

that will foster such critical thinking in the context of engineering. Ideally, students need to be introduced as early as their first year to metacognitive language and activities that allow for self-realization of their effective learning styles, so that when they are faced with a capstone design project, they will be able to practice skills rather than having to learn and apply at once.

Engineering students also frequently struggle with developing professional skills, and particularly with appreciating the value of those skills, which they might classify as touchy-feely or soft skills, compared to the more technical competencies that have traditionally been associated with engineering (Shuman, Besterfield-Sacre, & McGourty, 2005). Engineering students often self-select based on their technical skills, not their interpersonal skills, and thus instilling in them the value of nontechnical skills requires reinforcement throughout the curriculum.

FRAMEWORKS FOR DISCIPLINED REFLECTION BY STUDENTS

Christine Hogan (1995) proposed a structured journal writing activity based on the acronym SAID (Situation, Affect, Interpretation, Decision). It is a step-wise approach whereby the students document the following:

Situation: What actually happened?

- What images/scenes do you recall?
- Which people/words/comments struck you?
- What sounds/smells/sensations do you recall?
- Were there any other elements?

Affect: Incorporating your feelings and intuitions is important.

- What was the high/low spot?
- What was your mood/feeling?
- What was your gut reaction?

Interpretation: What did you learn?

- What can you conclude from this experience?
- What was your learning?
- How does this relate to appropriate concepts, theories, skills?

Decision: What will you do as a result?

- What do you need to do before this sort of thing happens again?
- What should you do differently next time?
- What would you say to people who weren't there?
- What was the significance of this experience in your life?

The SAID framework has been demonstrated to be an effective tool to guide engineering students in disciplined reflection in order to extract the lessons learned from projects and other practice-based learning experiences. (Jolly & Radcliffe, 2000; Walther et al., 2009).

Another approach to guiding students toward a disciplined approach to reflecting and thereby capturing transferable lessons learned from one design project and applying these to the next one is the SII method (Wasserman & Beyerlein, 2004). SII stands for Strengths, areas for Improvement, and Insights.

(S) Strengths: Identify the ways in which a performance was of high quality and commendable. Each strength statement should address what was valuable in the performance, why this attribute is important, and how to reproduce this aspect of the performance.

Improve Processes

(I) Areas for Improvement: Identify the changes that can be made in the future, between this assessment and the next assessment, that are likely to improve performance. Improvements should recognize the issues that caused any problems and mention how changes could be implemented to resolve these difficulties.

(I) Insights: Identify new and significant discoveries/understandings that were gained concerning the performance area—for example, what did the assessor learn that others might benefit from hearing or knowing? Insights include why a discovery or new understanding is important or significant and how it can be applied to other situations.

There are numerous other frameworks in the literature that provide a structured basis for disciplined reflection. One advantage of methods like SAID over that of SII is that the former method pivots on getting at the emotions (affect), how it felt for the students. Often the best reflections and the deepest learning comes from critical incidents or aha moments that are impactful to the individual because of the visceral impact of the event.

APPLICATION OF DISCIPLINED REFLECTION IN A DESIGN CLASS

It is widely recognized that assessment drives learning, or at the very least it focuses the attention of the student. Thus, asking students to reflect on and even self-evaluate their work at these times of assessment, summative or formative, has the emotional hook necessary. Each type of assessable task in a typical student design project affords unique opportunities for students to be asked to reflect and learn. This can be in relation to the technical work they

have done or to their teaming or other process skills in conducting a project.

Peer Reflection on Presentations

Immediately following a series of in-class presentations, it is helpful to ask each team to consider two questions:

1. What did you like or especially admire about the presentations of the other teams?
2. How might you adopt (and adapt) this to your next presentation?

This is best undertaken as a think-pair-share activity. Each team member takes a few minutes to write down as many ideas as they can to answer the questions. Then the team members share their ideas in pairs or as a whole team (depending upon the team size). Finally, there is a full-class discussion about the answers that each team decided upon. This helps affirm good ideas from other teams; peer recognition is a powerful incentive.

Reflection on Interim Team Reports

When projects are turned in to be graded there is a tendency for students to wait a week to get their grades back and then react. They can easily get upset when their visually stunning report, which they had spent an all-nighter to prepare, has lots of red ink on it with numerous comments and corrections. To avoid this type of reaction, and to foster self-assessment, one strategy is to hand back an unmarked copy of the report to each team member (on paper or electronically). Then, after reprising the lecture(s) given earlier in the course or the program on report writing, or the notes on report writing that they are meant to follow, the students are asked to spend 20 minutes individually

Improve Processes

reading and correcting their team report, especially the parts they wrote. Suddenly the errors and omissions will become all too obvious. Then, students are asked to share with their team what each found in the way of typos and spelling or grammar errors, as well as technical errors, poor word choice, inconsistencies, and so forth. The class is then asked to suggest a grade for the work based on the rubric that was made available before the submission was due and which was used to grade the report.

Now that students have calibrated on what was expected and have looked at their work through a fresh set of eyes, a week or so removed from the frantic rush to complete the report, the corrected and graded reports can be distributed. Now they are prepared to see the feedback, and it is not so easy for them to think that the instructor or grader was being harsh. Many lessons are driven home as a result. If this is done for an interim or preliminary (mid-semester) report, then the final reports are often significantly improved. A flexible grading system can also be used to measure the improvement, and thereby reward this learning.

Reflection on Design Processes

One method to encourage reflection on the design process as well as the technical outcome is to assign a substantial proportion of the report grade to be based on a critical reflection on team processes. There are a number of facets of this with relevant trigger questions. In each case the team is required to address the question and in making their case to draw upon evidence gathered during the course of the project. The sources of this evidence might include such things as team meeting minutes, document trails that illustrate the stages of the work, notes from meetings with stakeholders, and changes in

documents including task description, scope, requirements, and so forth.

In an interim or preliminary report, the sorts of process topics to be reported might include the following:

A critical analysis of team processes: What team tools were used, when, why, and what happened. Arguments are to be supported with evidence.

Lessons learned: The major lessons the team has learned through the process thus far. This might be related to organization, interactions, team interdependence, communication, performance, or other critical aspects of how the team got the work done. Arguments are to be supported with evidence.

Process improvements: What the team is planning to do differently in the next phase of this project and why. What actions the team is going to take to improve performance, what they expect to result, and why they expect this.

Project management plan: How the team plans to manage the remainder of the project, including a detailed Gantt chart of the major tasks to be completed and any dependencies between these. The team is to justify these tasks, estimate how many person-hours each requires, and identify who is going to be assigned to each task.

A corresponding assessment rubric is shown in Table 14.1.

The quality of the documentation in an interim report provides an opportunity to give feedback on aspects of the information and knowledge management process that commenced when the project was set up (see Chapter 6). Some of the main criteria are logical structure; easy to read layout; effective use of diagrams; absence of errors; consistency; referencing of sources; effective use of appendices in

TABLE 14.1   *Interim Report Assessment Rubric for Process*

| | Standard | | | |
Criterion	Unacceptable	Marginal	Acceptable	Superior
Critical analysis of the team processes	No ability to work together productively/ professionally	Significant team problems in leadership, cooperation, and interaction	Leadership, cooperation, and interaction are all evident and acceptable	Utilized strengths of each team member fully
Lessons learned	Not done or done incorrectly	Incomplete or partially incorrect evaluation	Sound evaluation of processes with supporting evidence	Insightful/correct evaluation with strong supporting evidence
Process improvements	No ability to work together productively/ professionally	Significant team problems in leadership, cooperation, and interaction	Leadership, cooperation, and interaction are all evident and acceptable	Utilized strengths of each team member fully
PM plan for delivering the project	Not organized; did not meet deadlines	Difficulty converting goals into tasks; routinely missed deadlines	Identified tasks, but struggled with priorities and planning; missed few deadlines	Effectively organized, prioritized, and met deadlines

relation to body of report for including details, info sources, and so forth. A possible assessment rubric is shown in Table 14.2.

In a final design report the sorts of process topics to be reported might include the following:

Stakeholder interactions/information gathering: Students critically analyze issues around gathering and analyzing information and/or working with stakeholders. Based on this analysis, they propose strategies they will use in future design projects and explain why these strategies will overcome issues.

Evolution of scope: Students critically analyze the evolution of the project scope. Based on this

analysis, they propose what strategies they will employ to manage the scope of future design projects and explain why these will work.

Effective team processes: Students critically analyze one or more team processes, tools, or techniques that were *particularly effective*. They explain why it worked and propose ways to improve upon it in future projects.

Ineffective team processes: Students critically analyze one or more of the processes that *did not work well* in their team. They describe what attempts the team made to overcome the problem and what resulted. Based on this analysis, students propose what they will do differently in the future to avoid this problem.

Improve Processes

TABLE 14.2 *Interim Report Assessment Rubric for Communication of Information/Knowledge*

	Standard			
Criterion	Unacceptable	Marginal	Acceptable	Superior
Scope and focus of paper	Purpose unclear; no clear structure	Purpose stated, but not helpful; difficult to follow because of lack of continuity	Purpose clearly stated and helps structure work; logical format for information helps reader	Purpose clear and explains work structure; information presented logically and is interesting
Appropriate application of information	No grasp of information; not interpreted, or errors in interpretation	Major gaps in content; inappropriate content may be included	Appropriate choice of content; comfortable with content and can explain to some degree	Consistently appropriate content; full subject knowledge with full explanations and elaborations
Style/grammar	Numerous errors; not proofread	Several errors; needs thorough proofreading	A few minor errors	Almost perfect; a joy to grade
Documentation of sources	Although needed, none	Inadequate list; inconsistent citing and referencing	Minor reference problems; citing and referencing consistent	Complete, comprehensive list of references with consistent and logical system

An associated rubric is illustrated in Table 14.3.

Assessment of Forward Communication of Information and Knowledge

The amount of knowledge accumulated during the course of a design project is often very significant, even for a one-semester student project. The vast majority of this knowledge is lost when the team disperses after the project is over. A similar phenomenon happens in engineering projects in industry. While a widely recognized best practice is to maintain a lessons learned database with each project in engineering practice, this is honored more in the breach rather than in the observance. The operational reality is that the daily pressures of getting a project completed on time and on budget becomes an excuse for not capturing and recording lessons as they arise during the course of the project. Then there is a rush at the end of the project to populate the lessons learned database, but by then much has been forgotten and many personnel are focused on the next project.

Further, in engineering practice it is common for a project to last several years and for there to be many changes of personnel during

TABLE 14.3   *Final Report Assessment Rubric for Knowledge Management and Team Processes*

Criterion	Standard			
	Unacceptable	Marginal	Acceptable	Superior
Stakeholder interactions/ information gathering	Not done or done incorrectly	Incomplete or partially incorrect evaluation	Sound evaluation of processes with supporting evidence	Insightful/correct evaluation with strong supporting evidence
Evaluation of scope	Not organized; did not meet deadlines	Difficulty convert goals into tasks; routinely missed deadlines	Identified tasks, but struggled with priorities and planning; missed few deadlines	Effectively organized, prioritized, and met deadlines
Effective team processes	No ability to identify instances of how to work together productively/ professionally	Can identify but not reflect usefully upon team success in leadership, cooperation, and interaction	Can identify and reflect usefully upon team success in leadership, cooperation, and interaction	Can identify and demonstrate deep insights around team success in leadership, cooperation, and interaction
Ineffective team processes	No ability to identify instances of a team not working together productively/ professionally	Can identify but not reflect usefully on instances of a team not working together productively/ professionally	Can identify and reflect usefully on instances of a team not working together productively/ professionally	Can identify and demonstrate deep insights into instances of a team not working together productively/ professionally

the course of the project. Each time a new team member joins, that person has to come up to speed and ideally acquire the knowledge already accumulated in the team. Most engineers have experienced the frustration of picking up a project partway through it and trying to fill in the missing pieces of information and surmise the tacit knowledge needed to understand the incomplete documentation that they inherit from the earlier phase of a project.

So, the educational challenge is to have students prepare their final reports and the ac-

companying collection of data, calculations, and sundry other material in such a way that it would make sense to another team who is handed their report two years later and expected to take the project to the next stage. With this in mind there are two criteria that should form the basis of assessing how robust and future-proof the final student team report is: *completeness* and *quality*.

Completeness includes such items as a comprehensive collection of information used and sources (e.g., prior art including literature); all

Improve Processes

TABLE 14.4 *Final Report Assessment Rubric for Communication of Information/Knowledge*

Criterion	Standard			
	Unacceptable	Marginal	Acceptable	Superior
Completeness	No grasp of information; not interpreted or errors in interpretation	Major gaps in content; inappropriate content may be included	Appropriate choice of content; comfortable with content and can explain to some degree	Consistently appropriate content; full subject knowledge with full explanations and elaborations
Quality	Purpose unclear; no clear structure	Purpose stated, but not helpful; difficult to follow because of lack of continuity	Purpose clearly stated and helps structure work; logical format for information helps reader	Purpose clear and explains work structure; information presented logically and is interesting
	Numerous errors; not proofread	Several errors; needs thorough proofreading	A few minor errors	Almost perfect; a joy to read

the people contacted (details so others can follow up); the critical information, analysis, and engineering calculations and assumption that support the main technical report (this might include photocopies from workbooks or indexing of workbooks). The quality relates to how easy it is to navigate the document and thus the ability to pick the project up where it left off. This is influenced by the report structure; layout; effective use of figures, illustrations, tables, and charts; use of appropriate technical communication style; absence of spelling and grammar errors; consistency; thoroughness in referencing of sources; overall impression; effective use of appendices in relation to body of report for including details, information sources, and so forth.

A relevant assessment rubric is shown in Table 14.4.

SUMMARY

Engineering design is a learning process that not only consumes existing knowledge but which also generates new knowledge. This new knowledge can be technical or process oriented in nature. Failure to adequately identify, capture, and reuse this new knowledge can lead to reinventing of the wheel each time a new project is undertaken and possibly the repeating of past mistakes. Studies of engineering practice suggest that design teams are neither particularly diligent nor effective in acquiring or using this new knowledge. In order to develop these necessary and essential skills of reflecting on practice and thereby learning, we propose strategies that encourage and reward reflective behaviors in engineering students. These strat-

Improve Processes

egies are based on structured approaches that foster disciplined reflection, preferably based on the emotional impact of critical incidents in projects.

REFERENCES

Hogan, C. (1995). Creative and reflective journal processes. *The Learning Organization*, *2*(2), 4–17. http://dx.doi.org/10.1108/09696479510086208

Jolly, L., & Radcliffe, D. F. (2000). Strategies for developing reflexive habits in students. *Proceedings of the 2000 ASEE Annual Conference*. Washington, DC: American Society for Engineering Education.

National Research Council. (2005). *How students learn: Science in the classroom*. Washington, DC: The National Academies Press.

NASA. (2010). *NASA Engineering Network and NASA Technical Report Server*. Open Government Initiative. Retrieved from http://www.nasa.gov/open/plan/nen-ntrs.html

Office of Inspector General. (2012). *Review of NASA's lessons learned information system*. Report No. IG-12-012 (p. 15). Washington, DC: Author. Retrieved from http://oig.nasa.gov/audits/reports/FY12/IG-12-012.pdf

Shuman, L. J., Besterfield-Sacre, M., & McGourty, J. (2005). The ABET professional skills—Can they be taught? Can they be assessed? *Journal of Engineering Education*, *94*(1), 41–55. http://dx.doi.org/10.1002/j.2168-9830.2005.tb00828.x

Walther, J., Kellam, N. N., Radcliffe, D., & Boonchai, C. (2009). Integrating students' learning experiences through deliberate reflective practice. *Proceedings of the 39th ASEE/IEEE Frontiers in Education Conference*. Piscataway, NJ: IEEE Press.

Wasserman, J., & Beyerlein, S. (2004). *SII method for assessment reporting*. Pacific Crest Faculty Development Series. Retrieved from http://www.webpages.uidaho.edu/ele/scholars/practices/Assessment/Resources/SII_Method.pdf

Improve Processes

PART III

Ensuring That Students Develop Information Literacy Skills

INTRODUCTION

The previous chapters in this handbook outline the place of information literacy within engineering design. This chapter complements the other chapters by showing how instructors can lay a foundation for students so that their first exposure to using information in an engineering context is not when they are engaged in a fully autonomous design project. In this chapter methods are described for assessing information literacy and provide examples that help gradually build student knowledge and skills as early as the first year of the engineering curriculum.

This chapter starts with a review of common challenges faced by undergraduate engineering students. Understanding these challenges is necessary in guiding the development of targeted instruction. We also emphasize the need for ongoing assessment and feedback, which is integral to scaffolding student learning. The strategies we discuss are designed to support student learning while gradually reducing the instructor support as students become more competent and independent.

COMMON CHALLENGES FOR STUDENTS

Gaining an accurate measure of students' skill and ability levels is a longstanding problem within education. Methods of quickly obtaining measures of student learning are, by nature, likely to focus too heavily on shallow conceptual understanding or students' perception of learning, rather than their actual learning (Wiersma & Jurs, 1990). There are, however, often disparities between students' perceived actual skill levels. For example, despite the complexity of the behaviors and skills necessary for information literacy, novice engineering students often perceive their information literacy skills to be higher than their actual skills (Holliday & Li, 2004; Ross, Fosmire, Wertz, Cardella, & Purzer, 2011).

Students, however, are able to identify specific skills that they find challenging. For example, they find creating a plan of action and locating information efficiently to be their key challenges (Head & Eisenberg, 2009). These challenges are associated with information-seeking behavior. In addition, our observations of students' actual performance show common errors in the following areas:

- Selection of inappropriate, untrustworthy resources (evaluating)
- Incorrect calculations and incorrect representation of scientific facts and information (applying)
- Misuse of information through exaggeration of information or misrepresentation of data (applying)
- Inconsistent documentation of information sources and citation errors (documenting)

These errors are associated with four areas of information literacy that we summarize in a framework called the *InfoSEAD model*: information seeking, evaluating, applying, and documenting. The information literacy behaviors of seeking and evaluating information as well as documenting and applying resources are essential during design projects. Students' common errors and weaknesses in key aspects of information literacy influence the quality of their arguments. In addition, documentation and citation errors are concerning in other ways as well. First, inappropriate or inconsistent citations point to haphazard collection of resources and impact the face quality of student reports and similar documents. Second, the use of ex-

CHAPTER 15

SCAFFOLD AND ASSESS

Preparing Students to Be Informed Designers

Senay Purzer, Purdue University

Ruth Wertz, Purdue University

Learning Objectives

So that you can actively promote the effective development of information literacy skills in student design teams, upon reading this chapter should be able to

- Explain common student challenges in information literacy

- Use assessments of information literacy for diagnostic purposes

- Use the InfoSEAD rubric for ongoing formative assessment and to provide feedback

- Implement scaffolding activities appropriate to students' information literacy skill levels and remove these scaffolds when appropriate

ternal information without appropriate citation is a violation of academic and professional integrity and can have significant consequences and even legal complications.

Educators are faced with the need to address these student challenges in a context where students feel confident about their skills. Ongoing classroom assessments and feedback are needed to identify skill areas that need the most improvement along with carefully developed scaffolding activities that can help correct student perceptions while building their knowledge and skills.

THE INFOSEAD MODEL

Information literacy can be seen as a skill emerging from a combination of self-directed learning and reflective judgment (Wertz, Purzer, Fosmire, & Cardella, in press). This means that an information-literate individual should

not only be able to plan and pursue information searches but also have the skills necessary to evaluate the accuracy of information and the quality of information sources (ACRL, 2000). We organized this knowledge and these skills in a four-dimensional framework called InfoSEAD (Wertz, Purzer, et al., 2013), summarized in Figure 15.1. We present this model to the students in our first-year engineering course as an intuitive mnemonic way to internalize the core tenets of information literacy. Breaking down the Association of College and Research Libraries (ACRL) standards into language more convenient for students removes the jargon barrier that some information literacy instruction can pose to incoming students.

SEEKING: Where Do I Find Information?

The InfoSEAD model starts with seeking activity, which refers to the search for information from a variety of information sources. The

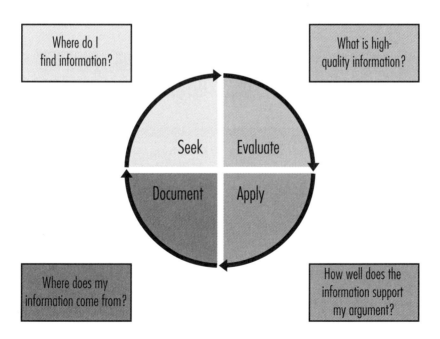

FIGURE 15.1 Four facets of information literacy in the InfoSEAD model.

search for information has to begin with a well-formed research question. Once students know what they are looking for, they then need to search in appropriate places to fill that information need. Examples of variety in information seeking are resources such as monographs, periodicals, and websites. The sources or authors of information can be internal or external to the organization. Some, such as conversations with peers, may also be more informal than the others but may play a critical role in the process.

EVALUATING: What Is High-Quality Information?

Once information sources are found and pieces of information identified, these need to be evaluated. Evaluation skills include the ability to critically evaluate the arguments made by the authors and identify the trustworthiness of the sources and references the arguments are based upon. These decisions can be made on the basis of the information source or the content of the material. The intended audience, such as popular or scholarly, can be an indicator of quality. Popular sources, though they are written for the general public and provide nonscientific or nontechnical information, can be appropriate in situations such as when the perceptions of users are sought. So, the evaluation of the quality of information depends on the context or situation.

APPLYING: How Well Does the Information Support My Argument?

Once information is evaluated and selected, it needs to be applied to the given situation and used to support design decisions. Information might be of high quality, but it also needs to be relevant to the situation under consideration. Students also need to be open to changing their

decisions or perspectives based on new information, rather than disregarding information that doesn't fit their hypothesis or misrepresenting the information contained in a document just to further their argument.

DOCUMENTING: Where Does My Information Come From?

The documentation of information sources is critical in several ways. First, documentation allows readers to judge the quality of information sources and hence the decisions made. Second, documentation acknowledges the sources cited and makes the document useful for those who may build on the information provided. Missing elements in a citation or reference make it difficult to link the information thread to the original source. Documentation errors can be as simple as citing and referencing errors or more substantial such as incorrect interpretation of information. Through in-text references arguments can be supported.

SCAFFOLDING STUDENTS' INFORMATION LITERACY SKILLS

In educational research, scaffolding is a metaphor used to describe temporary support provided to learners. Such support allows students to accomplish tasks that they are not able to accomplish otherwise (van de Pol, Volman, & Beishuizen, 2010). There are three critical characteristics of effective scaffolding: ongoing diagnosis, calibrated support, and fading. Scaffolding starts with a *diagnostic assessment* of student knowledge and skills. This diagnosis leads to the development of contingent or *calibrated support* appropriate for the needs of the learners. This support is then gradually reduced (i.e.,

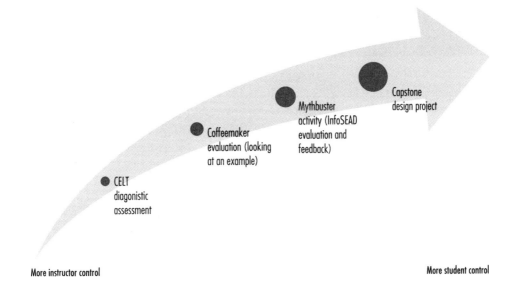

	CELT Diagnostic	Coffeemaker Activity	Mythbuster Activity	Capstone Design Project
Instructor Responsibility	Administer diagnostic assessment; identify weakness	Present clear examples; ask probing questions	Ask reflection questions	Ask broad guiding questions; provide feedback
Student Responsibility	Complete diagnostic assessment	Review and evaluate the example; identify exemplary components	Apply information skills in a well-defined in-class activity	Apply information skills in an ill-defined independent design process

FIGURE 15.2   Scaffolding process.

faded) as the learners become more competent in the task. Figure 15.2 demonstrates a scaffolding process involving two scaffolding activities that starts with a diagnostic assessment and gradually transfers responsibilities from the instructor to the students.

Diagnostic Assessment

Because effective scaffolding requires differentiated support, the process starts by assessing student learning and skills associated with a given task. Our recommendation is to start with an easy to administer and easy to score instrument for initial diagnosis. The Critical Engineering

Literacy Test (CELT) is an instrument developed for this purpose (Wertz, Saragih, et al., 2013). CELT is a multiple choice instrument and hence easy to administer and score. It starts with a text and a series of questions about this text. The full instrument is available upon request from the authors.

While CELT is administered once in our scaffolding process, ongoing assessments occur frequently through other formal or informal means to allow calibration of scaffolding.

Another form of scaffolding includes clarification of expectations. An evaluation rubric, shown in Table 15.1, provides characteristics of good quality outcomes and allows students to

TABLE 15.1 *InfoSEAD Assessment Rubric*

		Developing	Emerging	Proficient
Seek	Source quantity	Citations were fewer than the required quantity	Citations met the required quantity	Citations exceeded the required quantity
Evaluate	Source quality	Few sources are appropriate*	Most sources are appropriate*	All sources are appropriate*
Apply	Argument	Argument is disorganized with inconsistent use of evidence for support	Argument is understandable and somewhat supported with evidence	Argument is well structured and clearly supported with evidence
Document	Citations	Few citations are complete	Most citations are complete	All citations are complete and consistently formatted
	References	Few citations, tables, charts, and/or figures are referenced in text	Most citations, tables, charts, and/or figures are referenced in text	All citations, tables, charts, and/or figures are referenced in text
Subject-matter context	Subject literacy	Mostly incorrect use of terminology, scientific data, and units (several errors or misrepresentations)	Mostly correct use of terminology, scientific data, and units (a few minor errors)	Correct use of terminology, scientific data, and units

*Appropriate sources may include scholarly journals, technical reports, textbooks, and handbooks. Web resources such as government reports and product reviews may be acceptable but should be used only after careful assessment of the intended audience and purpose.

engage in self-evaluation as they develop their report. This InfoSEAD rubric can further be operationalized and familiarized to the students by having students evaluate the example report in Figure 15.3.

Calibrated Support

The results from CELT or a similar assessment should guide the development of calibrated instruction. Such instruction can take many forms ranging from modeling to questioning strategies. To scaffold student knowledge and

skills in information seeking and documentation, we provide a model report for students to analyze.

Coffeemaker Activity: Scaffolding by Modeling and Discussing a Written Example

This example report (Figure 15.3) models appropriate information documentation evidenced by in-text citations and a list of references and information seeking modeling the use of high-quality references, including peer-

Part I: Read the following narrative.

Evaluating the Design of a Coffeemaker

The objective of this report is to evaluate energy consumption associated with coffee making. Our analysis has shown that current coffeemaker machines are energy efficient and that the major energy cost occurs during the production of coffee.

According to the U.S. Department of Energy, the power requirements for coffeemakers range from 900 to 1200 watts. We conducted an experiment using a wattage measuring device, Kill-A-Watt, to test the power consumption of a Black & Decker coffeemaker. Our results showed that when the machine was turned on and the brewing cycle was started, the meter recorded a power consumption of 1 kilowatt hour (kWh). Assuming that the machine is used for one hour every day in a household and that electricity costs 10 cents per kWh, the cost of this machine's energy use would be 10 cents a day, or about 365 kWh annually. Assuming that all 115 million households in the U.S. (Day, 1996) use coffeemakers, the annual energy consumption for making coffee in the U.S. would be 42 × 109 kWh.

According to a research study conducted by Heller and Koelejan (2000), 10 percent of the energy used annually in the U.S. is consumed for producing food (based on data for 1994). Figure 1 shows the energy needed to produce a can of corn where the total energy input is 2.6 kWh. If all U.S. households consume one can of corn daily, the total energy need would be 111 × 109 kWh. Because we were not able to find data specifically for coffee production, we will assume that the energy needed for the production of coffee will be no less than the production of corn. The energy required to operate our individual coffeemaker (approximately 42 × 109 kWh) is significantly less than the energy used to process coffee (approximately 111 × 109 kWh per year). Therefore, we will focus on reducing the energy costs involved in producing coffee. Our boundary of analysis includes the production, processing, and packing of coffee beans and their transportation to and distribution in the mainland.

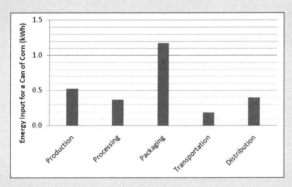

FIGURE 1 Energy input needed to produce a 455 g can of corn. (Modified from Pimentel & Pimentel, 1996.)

References

Day, J. C. (1996). *Projections of the number of households and families in the United States: 1995 to 2010*. U.S. Bureau of the Census, Current Population Reports, P25-1129. Washington, DC: U.S. Government Printing Office.

Heller, M. C., & Keoleian, G. A. (2000, December). *Life cycle-based sustainability indicators for assessment of the U.S. food system*. Report No. CSS00-04. Ann Arbor, MI: Center for Sustainable Systems, University of Michigan, 2000: 42. Retrieved from http://css.snre.umich.edu/css_doc/CSS00-04.pdf

Pimentel, D., & Pimentel M. (1996). Food processing, packaging, and preparation. In D. Pimentel, & M. Pimentel (Eds.), *Food, energy, and society* (revised ed., pp. 186–201). Niwot, CO: University Press of Colorado.

U.S. Department of Energy. (2010). *Estimating appliance and home electronic energy use*. Retrieved from http://www.energysavers.gov/your_home/appliances

Part II: Answer the following questions after reading the sample text.

InfoSEAD category	Reflection questions
Seeking	What three keywords might the authors of this report have used to find trustworthy information on this topic?
Evaluation	What aspects of this report help it make a strong argument?
Application	Give examples of how the authors apply information sources appropriately and inappropriately to their argument?
Documenting	How well have the authors documented their resources? What information still needs to be documented?

FIGURE 15.3   Coffeemaker scaffolding activity.

Mythbusters of Information

In this assignment, your task is to research a common belief and write an argument on. Please note that you will not conduct an experiment (or blow up stuff, as done in the popular Discovery Channel show *MythBusters*) to test the problem. Rather, you will conduct a literature search (e.g., search information using the library resources) to justify your arguments.

- You should cite at least four trustworthy external sources.
- Use in-text citations to support your arguments. In other words, show how your external information sources support your statements.
- Use correct terminology, scientific information, etc.
- Provide a clear and coherent argument.
- All citations should be in APA format.

Select one from the following statements/common beliefs:

- The carbon footprint of electrical cars is smaller than that of a comparable conventional gasoline-powered vehicle.
- Frozen vegetables are less nutritious than fresh vegetables.
- Cell phones that are left on could cause an airplane to crash.
- A person sitting in a car will not be hurt if the car is struck by lightning.

Suggested outline/structure

- First paragraph: What is the issue (claim)?
- Second paragraph: What are the reasons? What is the evidence and reasoning?
- Third paragraph: What are the counter arguments? Rebuttal?
- Fourth paragraph: What are the conclusions?
- References

FIGURE 15.4 Mythbuster scaffolding activity.

reviewed journal articles and textbooks. The instructor can further expand expectations for evaluating and applying by describing or speculating on the underlying decisions that led to the brief report on coffeemakers.

The example in Figure 15.3 is presented to the students along with a list of reflection questions that highlight key aspects of the report. The report models expected behaviors in referencing and in-text citation.

Mythbuster Activity: Scaffolding by Application and Feedback

The mythbuster activity (Figure 15.4) is structured as a team or a pair activity that can be done in the classroom, assuming students have access to the Internet to conduct their research. After students complete their report, they can be provided with feedback through instructor evaluation, peer evaluation, or self-evaluation using the InfoSEAD rubric.

Fading Support, Transferring Responsibilities

While the sample report on coffee making is a highly instructor-led activity, effective scaffolding requires the transfer of responsibilities from the instructor to the student over time in response to learning growth. The mythbuster assignment is an example of fading scaffolding that allows instructors to transfer responsibilities to the students so that they can engage in information evaluation and application. The scaffolding in this case is the InfoSEAD rubric that students are asked to follow as they conduct their research.

The scaffolding of information literacy is further removed as students engage in their design projects. Now they can take ownership and responsibility as they seek information from trustworthy resources, evaluate the quality and appropriateness of this information, apply this information to their design project, and correctly document their information sources. Prior to a capstone design project, instructors should reinforce the InfoSEAD approach throughout the engineering curriculum through the incorporation of mini-research papers, feasibility studies, and similar projects. Student mastery and internalization of the InfoSEAD (or similar) approach to information literacy will foster the increasingly independent, self-regulated learning that students will need to become effective lifelong learners throughout their post-graduate career.

SUMMARY

In this chapter we provided examples of on-going assessment tools and sample scaffolding activities that can help correct students' perceived beliefs about information literacy. These activities also support further development of students' information literacy skills. We also provided tools for the assessment of information literacy and hands-on application of these tools.

The scaffolding activities discussed in this chapter allow increasing levels of student competence and confidence in their information literacy skills. Lifelong learning can be achieved with necessary information literacy skills, as well as motivation and self-regulation. Hence, it is important to provide students with support that will lead to increased control over their learning.

ACKNOWLEDGMENTS

Aspects of this work received financial support from the Purdue College of Engineering's Engineer of 2020 Seed Grant program.

REFERENCES

ACRL. (2000). *Information literacy competency standards for higher education.* Chicago: American Library Association. Retrieved from http://www.acrl.org/ala/mgrps/divs/acrl/standards/standards.pdf

Head, A. J., & Eisenberg, M. B. (2009, December). *Lessons learned: How college students seek information in the digital age.* Project Information Literacy Progress Report, The Information School, University of Washington. Retrieved from http://projectinfolit.org/pdfs/PIL_Fall2009_finalv_YR1_12_2009v2.pdf

Holliday, W., & Li, Q. (2004). Understanding the millennials: Updating our knowledge about students. *Reference Services Review, 32*(4), 356–366.

Ross, M. C., Fosmire, M., Wertz, R. E. H., Cardella, M. E., & Purzer, S. (2011). Lifelong learning and information literacy skills and the first year engineering undergraduate: Report of a self-assessment. In *Proceedings of the 118th ASEE Annual Conference & Exposition.* Washington, DC: American Society for Engineering Education.

Van de Pol, J., Volman, M., & Beishuizen, J. (2010). Scaffolding in teacher–student interaction: A decade of research. *Educational Psychology Review, 22*(3), 271–296. http://dx.doi.org/10.1007/s10648-010-9127-6

Wertz, R. E. H., Purzer, S., Fosmire, M. J., & Cardella, M. E. (in press). Assessing information literacy skills demonstrated in an engineering design task. *Journal of Engineering Education.*

Wertz, R. E. H., Saragih, A., Van Epps, A. S., Sapp Nelson, M., Purzer, S., Fosmire, M. J., & Dillman, B. (2013). Work in progress: Critical thinking and information literacy: Assessing student performance. In *Proceedings of the 120th ASEE Conference & Exposition.* Washington, DC: American Society for Engineering Education.

Wiersma, W., & Jurs, S. G. (1990). *Educational measurement and testing.* Boston, MA: Allyn and Bacon.

CONCLUSION

We hope that this exploration of Information-Rich Engineering Design has sparked ideas that you will incorporate in your design classes to enable your students to make more effective use of a diverse range of information resources in their projects.

An informed approach to engineering design starts with laying a firm foundation, setting expectations for information gathering, and having teams develop codes of conduct for participating in information gathering and sharing resources among team members. Embedding the need for good information habits in the context of the ethical responsibilities of engineers, one of which is to provide accurate advice to clients, will impress upon students the need to take an informed approach seriously.

In the problem definition stage of the design process, students who uncover vital information well beyond that given to them by the client will produce more robust solutions—solutions more responsive to their clients' real needs. If students are guided to take the time to consider the solution context, environment, and culture they are designing for, and if their solutions meet professionally recognized external standards of performance, then they are becoming good engineering designers.

When synthesizing solutions, students who harness the substantial amount of prior art—knowledge of stuff that already exists—rather than attempting to reinvent it themselves, will save time, reduce costs, and come up with more sophisticated solutions with superior performance. By utilizing the information they've gathered within an evaluative structure, students will rapidly converge on the most promising solutions, thereby not wasting precious course time following false leads. By systematically analyzing materials and components, students similarly will efficiently locate the best materials for the job, rather than making do with suboptimal materials that may not be suited for the environment in which their design solution will be used.

Finally, students who manage their information effectively and efficiently will be able to draw upon it in the final documentation of their design project, providing just the information needed to make a persuasive, complete argument for their particular solution over other choices. And, once the project solution has been communicated, informed learners will reflect on their experiences in order to improve their professional practice, so they won't have to reinvent their own wheels in subsequent work.

In terms of implementing an information-rich approach to engineering design, we offer two practical pieces of advice. First, it is often easiest to implement information activities gradually over time. It is best to focus on one stage of the design process and to try implementing one of the activities or exercises suggested in this handbook, see what happens, improve, and iterate. Completely overhauling a course can be a way to make a clean break with past activities, but if the instructors and students are trying to master a new approach at the same time, the results can be disorienting and frustrating for both, and the new approach abandoned without being given a full test.

Second, if you value the information activities, make sure the course grades reflect that emphasis. Students are typically strategic learners. If they see that the bibliography of their reports is only worth five points, they will devote five points' worth of effort to gathering information. Providing positive reinforcement throughout the course that information is important and expecting them to gather information at different stages of their design process, on the other hand, will help students internalize that ethos, and the practice will make it easier to locate information in their future activities.

This process works best when engineering educators and librarians work together as a team. Librarians will be aware of the latest information tools and resources, best practices in information organization, and how to extract relevant and appropriate information from technical sources. Engineering educators understand the design process and will have an intuitive feel for the challenges students face and the pedagogies that resonate with them. They will be more familiar with the content of technical information and can share how they use information in their own practice. Integrating the synergistic strengths of these two professionals can transform the ways engineering design is taught and how information literacy is acquired by students.

If you are an engineering educator, we recommend that you find your institution's librarian and see how you can work together to make students aware of all the resources available to them, and guide them in how to locate, evaluate, and apply that information to their design projects. Higher quality projects are much less onerous to grade, so time invested in teaching information skills to students will reap rewards at the end of the course. If you are a member of the American Society for Engineering Education (ASEE), check out the activities of the Engineering Librarians Division at the annual conference.

If you are a librarian, track down the engineering design instructors at your institution and ask them about their course and what challenges seem most difficult for students, and see if any of them resonate with some of the ideas discussed in this handbook. If so, you can suggest that they try some activities to help students meet those challenges. Small successes can lead to more substantial collaborations, and eventually, perhaps, to a full-blown information-rich design process. Remember that design activities may be taking place across the engineering curriculum, from a first-year introduction course to a capstone design experience. Some engineering programs are experimenting with incorporating a "design spine" where the students have a structured design experience each year, if not each semester. Students are afforded the chance to build increasingly sophisticated information skills if they are embedded sequentially across the curriculum in a purposeful manner.

Our hope is that sharing this handbook with your counterpart at your institution will lead you to productive discussions and potential collaborations to help your students learn professional skills in an authentic design context. Ultimately, we believe that taking an information-rich approach to engineering design will lead to students better able to function and stay abreast of innovations in our fast-moving modern engineering profession.

CONTRIBUTORS

Jay Bhatt is the liaison librarian for the College of Engineering at Drexel University. He is responsible for building library collections in engineering subject areas, outreach to faculty and students, and teaching information and research skills to faculty and students in engineering, biomedical engineering, and related subject areas. He provides individual and small group consultations to students, instructional sessions to specific classes, online research support in both face-to-face and distance learning programs, and workshops for specialized research areas. Mr. Bhatt has published and presented papers extensively in the area of information literacy for engineering students.

Dr. Patrice Buzzanell is a professor of communication in the Brian Lamb School of Communication (and a professor of engineering education by courtesy) at Purdue University. Dr. Buzzanell is the author of 3 edited books and over 130 articles and chapters. Her research centers on the everyday negotiations and structures that produce and are produced by the intersections of career, gender, and communication, particularly in STEM (science, technology, engineering, and math).

Dr. Monica Cardella is an associate professor of engineering education and is the director of informal learning environments research for the Institute for P-12 Engineering Research and Learning (INSPIRE) at Purdue University. She received her MS and PhD degrees in industrial engineering at the University of Washington and her BS degree in mathematics from the University of Puget Sound. Dr. Cardella teaches and has served as a course coordinator in the first-year engineering program at Purdue, where she has tried out many of the approaches described in this book. Her current research focuses on the development of engineering thinking (primarily focused on design thinking and mathematical thinking) across the life span (i.e., from age four years through professional practice) in both formal and informal environments.

Jim Clarke earned a BA in history and communications from Hiram College, an MA in American history from the University of Houston, and an MLS from the University of Michigan. Mr. Clarke has worked as an engineering librarian and as a product information manager for companies such as Ford Motor Company and International Truck and Engine Corporation, and within divisions of the DaimlerChrysler Truck Group. He currently is the engineering librarian for Miami University.

Donna Ferullo is the director of the University Copyright Office and associate professor of library science at Purdue University. She advises the university on copyright compliance issues and educates the Purdue University community on their rights and responsibilities under the copyright law. Ms. Ferullo holds a JD degree from Suffolk University Law School, an MLS degree from the University of Maryland, and a BA degree in Communications from Boston College. Ms. Ferullo has published articles on copyright and its impact on higher education and libraries, is past chair of the Association of College and Research Libraries' Copyright Committee, and serves on the copyright committee of the Indiana Partnership for Statewide Education (IPSE).

Michael Fosmire is the head of the physical sciences, engineering, and technology divisions and professor of library science of the Purdue University Libraries. He has written extensively on the role of information in active-learning pedagogies and the integration of information literacy in science and technology curricula and is the author of the *Sudden Selector's Guide to Physics*. He has also edited the physics section of the American Library Association's *Guide to Reference and Resources for College Libraries*.

Jeremy Garritano is an associate professor of library science and has been the chemical information specialist for the Purdue University Libraries since 2004, where he is the Libraries liaison to the areas of chemistry, chemical engineering, and materials engineering. Mr. Garritano holds a BS degree in chemical engineering from Purdue University and an MLS degree from Indiana University. His research interests include chemical information literacy and liaison librarian experiences with data management. Previously he has worked at George Mason University and Earlham College.

Jon Jeffryes is an engineering librarian at the University of Minnesota where he is subject liaison to the Departments of Biomedical, Civil, Industrial, and Mechanical Engineering and manages the Libraries Standards Collection. Mr. Jeffryes holds an MA-LIS degree from the University of Wisconsin-Madison and a BA degree in English from Grinnell College. His research interests are focused on the information needs of engineers and information literacy and teaching.

Michael Magee is a '14 year student at Drexel University studying architectural engineering with a mechanical concentration and a special emphasis in sustainable HVAC applications. He has been vice president for Drexel Smart House since spring 2010, and since 2009 he has been researching with the DSH Lightweight Green Roof team, which received the EPA P3 phase II award in 2011. Mr. Magee has been involved in several LEED projects during his past co-op positions, has completed a Passive House Planning Package (PHPP) energy model for a Habitat for Humanity feasibility study, and has assisted with the development of building energy and ventilation models associated with NIST's Net-Zero Energy Residential

Test Facility (NZERTF) in Gaithersburg, Maryland. He is dedicated and maintains a passion for the innovation and creativity required to push the new paradigm of responsible building practice in order to improve the quality of the built environment for our future.

Dr. Joseph Mullin is the Teaching Professor in the Civil, Architectural, and Environmental Engineering Department at Drexel University. Dr. Mullin received both his BS and MS degrees in civil engineering from Drexel University and his PhD degree from The Pennsylvania State University. His early research areas included biaxial fatigue studies on high performance aluminum alloys for aircraft. Later, at General Electric Space Sciences Lab, he was involved in developing composite materials for aerospace applications including heat shields for reentry systems and carbon epoxy structural members for spacecraft. He has also been teaching materials and structural courses at both the graduate and undergraduate level for many years with emphasis on failure mechanisms. His responsibilities include advising civil engineering senior design groups on structures, materials selection, and design optimization.

Megan Sapp Nelson is an associate professor of library science at Purdue University. Ms. Sapp Nelson holds MLS and BA degrees from the University of Illinois, Urbana-Champaign. She currently serves as liaison to the Schools of Civil Engineering, Construction and Engineering Management, Electrical and Computer Engineering, and Environmental and Ecological Engineering, as well as the Departments of Earth, Atmospheric, and Planetary Sciences and Electrical and Computer Engineering Technology. Her teaching and research focuses on information literacy–related professional skills needed by STEM students, including data information literacy, data management, and embedding information literacy into the engineering design cycle.

Dr. Bonnie Osif is an engineering reference and instruction librarian in the Engineering Library at The Pennsylvania State University. She holds a BS degree in biology from Penn State, an MS degree in information science from Drexel, and an EdD degree in science education from Temple University. She is active in the Special Libraries Association and the Transportation Research Board. She is the co-author of *TMI: 25 Years Later* and editor of *Using the Engineering Literature*. Dr. Osif has authored more than 100 papers and presentations.

Dr. Senay Purzer is an assistant professor in the School of Engineering Education and is the co-director of assessment research for the Institute for P-12 Engineering Research and Learning (INSPIRE) at Purdue University. Dr. Purzer received her MA and PhD degrees in science education at Arizona State University. She also holds a BS degree in physics education and a BSE degree in engineering. She has written journal publications on teaming and design, conceptual learning, and instrument development. Her current research focuses on design problem solving, assessment of lifelong learning, and K-12 engineering education.

Dr. David Radcliffe is the Kamyar Haghighi head of the School of Engineering Education and the epistemology professor of engineering education at Purdue University. He holds BEng and MEngSci degrees in mechanical engineering from the University of Queensland and a PhD in biomedical engineering from Strathclyde University. His teaching and research interests span engineering design, systems engineering, engineering education and professional development, innovative learning spaces, and knowledge management.

Amy Van Epps is an associate professor of library science and engineering librarian at Purdue University. Ms. Van Epps received an MSLS degree from the Catholic University of America, an MEng (IE) degree from Rensselaer Polytechnic Institute, and a BA degree in engineering science from Lafayette College. She has extensive experience providing instruction for engineering and technology students, including those in Purdue's first-year engineering program. Her research interests include finding effective methods for integrating information literacy knowledge into the undergraduate engineering curriculum.

Ruth Wertz is a doctoral candidate in the School of Engineering Education at Purdue University. She holds an MS degree in civil engineering from Purdue University and a BS degree in civil engineering from Trine University (formally Tri-State University). Ms. Wertz is a licensed professional engineer in the State of Indiana with over six years of field experience and eight years of classroom teaching experience. Her research interests include teaching and learning engineering in online course formats and the development of information literacy in engineering students.

Dr. Carla Zoltowski is co-director of the EPICS Program at Purdue University. She holds BSEE, MSEE, and PhD degrees in engineering education, all from Purdue, and is responsible for teaching design and developing curriculum and assessment tools for the EPICS Program. Dr. Zoltowski's academic and research interests include human-centered design, ethical reasoning, leadership, service learning, and assistive technology. She oversees the research efforts within EPICS.

INDEX

Page numbers in italics refer to tables, boxes, and figures.

Page numbers in italics refer to tables, boxes, and figures.

Page numbers in italics refer to tables, boxes, and figures.

Page numbers in italics refer to tables, boxes, and figures.

Page numbers in italics refer to tables, boxes, and figures.

Page numbers in italics refer to tables, boxes, and figures.